Lecture Notes in Mathematics

Edited by A. Dold and B. Eckmann

631

Numerical Treatment of Differential Equations

Proceedings of a Conference,
Held at Oberwolfach,
July 4–10, 1976

Edited by
R. Bulirsch, R. D. Grigorieff, and J. Schröder

Springer-Verlag
Berlin Heidelberg New York 1978

Editors

R. Bulirsch
Institut für Mathematik
der Technischen Universität
Arcisstraße 21
8000 München 2/BRD

R. D. Grigorieff
Technische Universität Berlin
Fachbereich Mathematik
Straße des 17. Juni 135
1000 Berlin 12/BRD

J. Schröder
Mathematisches Institut
der Universität zu Köln
Weyertal 86–90
5000 Köln 41/BRD

AMS Subject Classifications (1970): 34A50, 35-04, 35A40, 65B05, 65LXX, 65NXX, 65Q05, 65R05

ISBN 3-540-08539-4 Springer-Verlag Berlin Heidelberg New York
ISBN 0-387-08539-4 Springer-Verlag New York Heidelberg Berlin

Printed in Germany

Printing and binding: Beltz Offsetdruck, Hemsbach/Bergstr.
2140/3140-543210

P R E F A C E

From July 4 to July 1o, 1976, a meeting on the *Numerical Treatment of Differential Equations* was held at the Mathematisches Forschungsinstitut Oberwolfach, with the emphasis on

1) *Initial value problems for ordinary differential equations,* and

2) *fast methods for solving difference equations for elliptic boundary value problems.*

Concerning ordinary differential equations, the topics included the numerical treatment of stiff equations, differential equations with time-lag argument, investigations of domains of stability, solutions of singular boundary value problems and others.

The talks on the solution of algebraic systems arising from the discretization of elliptic boundary value problems were concerned with direct and iterative methods, capacitance matrix methods, and the systematic testing of computer programs. In addition, several talks on various other subjects related to the above topics were given.

Since it is the belief of the organizers of this meeting that presentations of this meeting would be of some interest to a larger part of the numerical analysis community, we want to express our gratitude to the authors of the published contributions as well as to SPRINGER-Verlag who made this publication possible.

April 1976

München Köln Berlin

TABLE OF CONTENTS

---

*) Talks given, but not included in the volume

ADDRESSES OF THE AUTHORS

Prof. Dr. G. Alefeld

Institut f. Angew. Mathematik
der Universität Karlsruhe
Englerstr.

D - 7500 Karlsruhe

Prof. Dr. D.G. Bettis

Dept. of Aerospace Engineering
The University of Texas at Austin

Austin, Texas 78712 USA

Prof. Dr. B.L. Buzbee

University of California
Los Alamos Scientific Laboratory

Los Alamos, New Mexico 87544 USA

Prof. Dr. G. Dahlquist

Inst. för Informationsbehandling
Numerisk Analys
Kungliga Tekniska Högskolan

10044 Stockholm 70 Schweden

Dr. Reinhard Frank

Institut f. Numerische Mathematik
Technische Hochschule Wien
Gusshausstr. 27 - 29

A - 1040 Wien

Dr. A. Friedli

Eidgenössische
Technische Hochschule Zürich
Seminar f. Angew. Mathematik
Clausiusstr. 55

CH - 8006 Zürich

Dr. Wolfgang Hackbusch

Mathematisches Institut
der Universität zu Köln
Weyertal 88

D - 5000 Köln 41

Prof. Dr. Rolf Jeltsch

Abteilung für Mathematik
der Universität Bochum
Gebäude NA

D - 4630 Bochum - Querenburg

Dr. R. Mannshardt

Rechenzentrum
der Universität Bochum
Gebäude NA

D - 4630 Bochum - Querenburg

Dr. B. Mehri

Department of Mathematics
ARYA-MEHR University

Teheran - Iran

Prof. Dr. T. Meis

Mathematisches Institut
der Universität zu Köln
Weyertal 86 - 9o

D - 5000 Köln 41

Prof. Dr. K. Nickel

Institut f. Angewandte Mathematik
der Universität Freiburg
Hebelstr. 40

D - 7800 Freiburg / Brsg.

Civ.Ing. J. Oppelstrup

Royal Institute of Technology
Dept. for Computer Sciences
Numerical Analysis

S-10044 Stockholm 70 Schweden

Dr. R. Scherer

Mathematisches Institut
Universität Tübingen
Auf der Morgenstelle 10

D - 7400 Tübingen

Prof. Dr. J. Schröder

Institut für Mathematik
der Universität zu Köln
Weyertal 86 - 90

D - 5000 Köln 41

Prof. Dr. Hans Stetter

Institut f. Numerische Mathematik
Technische Hochschule Wien
Gusshausstr. 27 - 29

A - 1040 Wien

Dr. U. Trottenberg

Mathematisches Institut
der Universität zu Köln
Weyertal 86 - 90

D - 5000 Köln 41

Dr. J. Waldvogel

Eidgenössische Hochschule
Zürich
Seminar f. Angew. Mathematik
Clausiusstr. 55

CH - 8006 Zürich

Prof. Dr. O. Widlund

Courant Institute of
Mathematical Sciences
University Computing Center

New York, New York 10012 USA

Dr. K. Witsch

Mathematisches Institut
der Universität zu Köln
Weyertal 86 - 90

D - 5000 Köln 41

Some Convergence Results for the PEACEMAN-RACHFORD Method in the Noncommutative Case

G. Alefeld

1. Introduction

We consider the system of simultaneous linear equations

$$Au = b \ .$$

Let the matrix A be expressed as the sum

$$A = H + V$$

of two matrices H and V. Then we consider the following iteration method for solving the system given above:

$$\begin{cases} (r_k I + H)x_{k+\frac{1}{2}} = (r_k I - V)x_k + b \\ (r_k I + V)x_{k+1} = (r_k I - H)x_{k+\frac{1}{2}} + b \ , \end{cases}$$

$$k = 0,1,2,\ldots \ ,$$

$$(r_k > 0 \ , \quad I = \text{unit matrix}) \ .$$

This method is called PEACEMAN-RACHFORD iterative method (PRM). If $r_k = r$, $k = 0,1,2,\ldots$, then the method is called stationary otherwise nonstationary.

Most known results concerning the convergence of the stationary PRM consider the case in which both H and V are Hermitian and nonnegative definite and where at least one of the matrices H and V is positive definite ([13,14,16]).

In the nonstationary case very satisfactory practical experience has been made. But proofs of convergence and optimizing the parameter sequence (r_k) have been performed only under even more restrictive conditions ([13,14,16]). Particularly the matrices H and V have to commute, this means that $HV = VH$ holds. Although the nonstationary method shows very good convergence behavior also in most noncommutative cases there are scarcely criteria known which assure at least convergence in these cases. See however [2,6,7,11,15]. On the other side there are linear systems arising from boundary value problems for which it is possible to choose a parameter sequence (r_k) such that PRM does not converge ([12]). Because of these reasons it seems quite desirable to look for new convergence criteria for the non-stationary method.

In this paper we first report on some results from ALEFELD [1] concerning the convergence of PRM (Section 2). These results can immediately be applied to discrete versions of elliptic boundary value problems (Section 3). Finally we prove a new convergence result for an iterative method for $m \geqq 2$ space variables which was introduced in [5] (Section 4).

2. A Convergence Theorem for PRM

Let $\mathbb{C}^{n,n}$ be the set of all $n \times n$ matrices $A = [a_{ij}]$ with elements taken from $\mathbb{C}$ (= the set of complex numbers). Analogously $R^{n,n}$ is defined. For $A = [a_{ij}] \in \mathbb{C}^{n,n}$ we set $|A| := [|a_{ij}|] \in R^{n,n}$. Let $A = [a_{ij}] \in \mathbb{C}^{n,n}$ be decomposed into the sum $A = H + V$ and let

$$H = D_H - B_H \qquad \text{and} \qquad V = D_V - B_V .$$

Here D_H and D_V represent the diagonal parts of H and V whereas B_H and B_V stand for the off-diagonal parts of H and V respectively. The set of matrices $\tilde{\Omega}(A)$ is now defined by

$$\tilde{\Omega}(A) := \{C = [c_{ij}] \in \mathbb{C}^{n,n} \mid C = \tilde{H} + \tilde{V},\ \tilde{H} = D_{\tilde{H}} - B_{\tilde{H}},\ \tilde{V} = D_{\tilde{V}} - B_{\tilde{V}},$$

$$D_{\tilde{H}} = D_H, |B_{\tilde{H}}| = |B_H|,\ D_{\tilde{V}} = D_V,\ |B_{\tilde{V}}| = |B_V|\}.$$

We have $A \in \tilde{\Omega}(A)$.

The spectral radius of a matrix A is denoted by $\rho(A)$. Consider now any $B = [b_{ij}] \in R^{n,n}$ with $b_{ij} \leqq 0$, $i \neq j$. Then B can be expressed as the difference

$$B = \kappa I - C$$

where $\kappa = \max\limits_{1 \leqq i \leqq n}\{b_{ii}\}$ and where $C = [c_{ij}] \in R^{n,n}$, satisfying $C \geqq 0$, has its entries given by

$$c_{ii} = \kappa - b_{ii} \geqq 0 , \quad 1 \leqq i \leqq n$$

$$c_{ij} = -b_{ij} \geqq 0 , \quad i \neq j , \quad 1 \leqq i,j \leqq n .$$

Following OSTROWSKI [9] such a matrix B is called a nonsingular M-matrix iff $\kappa > \rho(C)$. By Theorem 3.8 and by Theorem 3.10 in [13] a nonsingular M-matrix has positive diagonal elements. The proof of the following theorem is given in detail in [1].

Theorem 1. Let the matrix $A = [a_{ij}] \in \mathbb{C}^{n,n}$ be expressed as the sum $A = H + V$ of two matrices $H = [h_{ij}]$ and $V = [v_{ij}]$. Let H and V both have only real diagonal elements. Let τ be defined by

$$\tau = \max_{1 \leqq i \leqq n} \{h_{ii}, v_{ii}\}$$

and assume that the matrices

$$\tau I + D_H - |B_H| \quad \text{and} \quad \tau I + D_V - |B_V|$$

are both nonsingular M-matrices. Then the following are equivalent:

(a) $D_V + D_H - (|B_V| + |B_H|)$ is a nonsingular M-matrix;

(b) For any matrix of the set $\tilde{\Omega}(A)$ and for any sequence (r_k) satisfying

$$\tau \leqq r_k \leqq \sigma < \infty, \quad k = 0,1,2,\dots,$$

($\sigma \geqq \tau$) PRM is convergent. ■

As a special case of Theorem 1 we get

Corollary 1. Let $A = [a_{ij}] \in R^{n,n}$ be decomposed into the sum $A = H + V$ of two real matrices $H = [h_{ij}]$ and $V = [v_{ij}]$. Let

$$\tau = \max_{1 \leqq i \leqq n} \{h_{ii}, v_{ii}\}$$

and assume that

$$\tau I + H \quad \text{and} \quad \tau I + V$$

are both nonsingular M-matrices. Then the following are equivalent:

(a) A is a nonsingular M-matrix;

(b) PRM is convergent for any matrix of the set $\tilde{\Omega}(A)$ and for any sequence (r_k) satisfying

$$\tau \leqq r_k \leqq \sigma < \infty, \quad k = 0,1,2,\dots,$$

($\sigma \geqq \tau$) . ■

3. Applications to Disctretized Elliptic Equations

Let R be a bounded plane region with boundary ∂R . Consider the linear second-order partial differential equation

$$L[u] \equiv Au_{xx} + Cu_{yy} + Du_x + Eu_y + Fu = G$$

with coefficients A,C,D,E,F and G which are functions of x and y and with $A \geqq m$, $C \geqq m$, $m > 0$ and $F \leqq 0$ in R . The function u is also required to satisfy the condition

$$u(x,y) = g(x,y)$$

on the boundary ∂R of R . Replacing the derivatives by difference quotients leads to a second-order partial difference operator

$$L_h[u] \equiv \alpha_0 u(x,y) - \alpha_1 u(x+h,y) - \alpha_2 u(x,y+h) - \alpha_3 u(x-h,y) - \alpha_4 u(x,y-h) = t(x,y)$$

where

$$\alpha_1 = A + \frac{h}{2} D , \qquad \alpha_2 = C + \frac{h}{2} E ,$$

$$\alpha_3 = A - \frac{h}{2} D , \qquad \alpha_4 = C - \frac{h}{2} E ,$$

$$\alpha_0 = \alpha_1 + \alpha_2 + \alpha_3 + \alpha_4 - h^2 F ,$$

$$t(x,y) = -h^2 G .$$

Here we have used the usual three-point central difference quotients. For simplicity we assume that it is not necessary to approximate the boundary ∂R . The equation $L_h[u] = t(x,y)$ is equivalent to a system of linear algebraic equations $Au = b$. It is well know that for

$$h < h_0 = \min\{\min_{R+\partial R} \frac{2A}{|D|} , \min_{R+\partial R} \frac{2C}{|E|}\}$$

the matrix A is a nonsingular M-matrix. Expressing $L_h[u]$ as

$$L_h[u] = H_h[u] + V_h[u]$$

where

$$H_h[u] = (2A - \tfrac{1}{2}h^2F)u(x,y) - (A + \tfrac{1}{2}hD)u(x + h,y) - (A - \tfrac{1}{2}hD)u(x - h,y),$$

$$V_h[u] = (2C - \tfrac{1}{2}h^2F)u(x,y) - (C + \tfrac{1}{2}hE)u(x,y + h) - (C - \tfrac{1}{2}hE)u(x,y - h)$$

the matrix A can be written in the form $A = H + V$ where H and V are both nonsingular M-matrices. But then the same is true for $H + \tau I$ and $V + \tau I$ ([9]). Therefore by applying Corollary 1 the following holds.

<u>Theorem 2.</u> <u>Let</u> $L_h[u] = -h^2 G$ <u>where</u> $h < h_0$.

Let

$$\tau = \max\{\max_{R+\partial R} (2A - \tfrac{1}{2}h^2F), \max_{R+\partial R} (2C - \tfrac{1}{2}h^2F)\}$$

Then PRM is convergent for any sequence (r_k) satisfying $\tau \leqq r_k \leqq \sigma < \infty$, $k = 0,1,2,\dots$, $(\sigma \geqq \tau)$. ∎

4. Remarks on Methods for $m \geqq 2$ Space Variables

We consider the problem of solving the system of linear equations

$$(A_1 + A_2 + \dots + A_m)x = b \quad , \qquad m \geqq 2 .$$

In [5] among others the following iterative method was proposed:

$$(rI + A_i)x_i^{(k+1)} = \sum_{j=1}^{i-1} (\tfrac{r}{m-1} I - A_j)x_j^{(k+1)} + \sum_{j=i+1}^{m} (\tfrac{r}{m-1} I - A_j)x_j^{(k)} + b, \qquad (*)$$

$$i = 1(1)m , \qquad k = 0,1,2,\dots, (r > 0) .$$

It was proved in [5] that provided the matrices A_i, $1 \leqq i \leqq m$, are alle Hermitian and positive definite and provided the eigenvalues $\lambda_j(i)$ of A_i satisfy $0 < a \leqq \lambda_j(i) \leqq b$, $1 \leqq i \leqq n$, $1 \leqq j \leqq m$, then for $r > (m-2)b/2$ it follows $\lim_{k\to\infty} x^{(k)} = z$, $i = 1(1)m$, where z is the solution of the given system. It was pointed out in [5] that it is important that this result holds without assuming commutativity of the matrices A_i. The same is true for the result given in the next theorem.

Theorem 3. *Let $A_i = [a_{st}^{(i)}]$, $1 \leqq i \leqq m$, be all nonsingular M-matrices. Then if A is a nonsingular M-matrix method (*) is convergent for*

$$r \geqq \tau := (m-1) \max_{\substack{1 \leqq s \leqq n \\ 1 \leqq i \leqq m}} \{a_{ss}^{(i)}\} .$$

Proof. Consider the $m\cdot n \times m\cdot n$ matrix $\tilde{A}_r$ given by

$$\tilde{A}_r = \begin{bmatrix} rI + A_1 & -(\tfrac{r}{m-1} I - A_2) & \dots & -(\tfrac{r}{m-1} I - A_m) \\ -(\tfrac{r}{m-1} I - A_1 & rI + A_2 & \dots & -(\tfrac{r}{m-1} I - A_m) \\ \dots & \dots & & \dots \\ -(\tfrac{r}{m-1} I - A_1) & -(\tfrac{r}{m-1} I - A_2) & \dots & rI + A_m \end{bmatrix}$$

and the splitting

$$\tilde{A}_r = \tilde{M}_r - \tilde{N}_r$$

where $\tilde{M}_r$ is given by

$$\tilde{M}_r = \begin{bmatrix} rI + A_1 & & & 0 \\ -(\frac{r}{m-1} I - A_1) & rI + A_2 & & \\ \cdots & \cdots & \ddots & \\ -(\frac{r}{m-1} i - A_1) & -(\frac{r}{m-1} I - A_2) & \cdots & rI + A_m \end{bmatrix} .$$

Defining the vectors $\bar{x} \in R^{m \cdot n}$ and $\bar{c} \in R^{m \cdot n}$ by

$$\bar{x} = \begin{bmatrix} x_1 \\ x_2 \\ \cdot \\ \cdot \\ \cdot \\ x_m \end{bmatrix}, \ x_i \in R^n , \quad i = 1(1)m , \quad \bar{c} = \begin{bmatrix} b \\ b \\ \cdot \\ \cdot \\ \cdot \\ b \end{bmatrix} , \quad b \in R^n ,$$

it is easy to see that (∗) can be written in the form

$$\bar{x}^{(k+1)} = \tilde{M}_r^{-1} \tilde{N}_r \bar{x}^{(k)} + \tilde{M}_r^{-1} \bar{c} , \quad k = 0,1,2,\ldots \ .$$

First we show that $\tilde{A}_r$ is a nonsingular M-matrix. To do this it is sufficient to verify that the off-diagonal elements of $\tilde{A}_r$ are nonpositive and that there exists a positive vector $\bar{z} \in R^{m \cdot n}$ with $\tilde{A}_r \bar{z} > \bar{o}$. (See [13], p.85 and exercise 2 on p. 87). Since by hypothesis the A_i are nonsingular M-matrices then the off-diagonal elements of $\tilde{A}_r$ are nonpositive if $r \geqq (m - 1)a_{ss}^{(i)}$, $1 \leqq s \leqq n$, $1 \leqq i \leqq m$. Again, since A itself is by hypothesis a nonsingular M-matrix there exists a positive vector $z \in R^n$ with $Az > 0$. If we take

$$\bar{z} = \begin{bmatrix} z \\ z \\ \cdot \\ \cdot \\ \cdot \\ z \end{bmatrix} \in R^{m \cdot n}$$

then a simple calculation shows $\tilde{A}_r \bar{z} > \bar{o}$. Hence $\tilde{A}_r$ is a nonsingular M-matrix. Especially, we have $\tilde{A}_r^{-1} \geqq 0$ ([13], p.85). Since any matrix obtained from a nonsingular M-Matrix by setting certain off-diagonal entries to zero,is also a nonsingular M-matrix we have that $\tilde{M}_r$ is a nonsingular M-matrix from which it follows again that $\tilde{M}_r^{-1} \geqq 0$. Furthermore $\tilde{N}_r \geqq 0$. Hence the splitting $\tilde{A}_r = \tilde{M}_r - \tilde{N}_r$ is a regular splitting ([13], Definition 3.5) and therefore by Theorem 3.13 in [13] the method (*) is convergent. This completes the proof of this theorem. ■

We close this paper with two remarks:

1. It is easy to give similiar convergence results for the other methods proposed in [5].

2. Theorem 3 can directly be applied to discrete versions of boundary value problems for $m \geqq 2$ space variables.

References

1. Alefeld, G.: Zur Konvergenz des Peaceman-Rachford-Verfahrens. To appear in Numer. Math.

2. Birkoff, G., Varga, R.S.: Implicit Alternating Direction Methods. Trans. Amer. Math. Soc. 92, 190-273 (1959)

3. Birkoff, G., Varga, R.S., Young, D.M.: Alternating Direction Implicit Methods. In: Advances in Computers 3, New York: Academic Press 1962

4. Casper, J.: Applications of Alternating Direction Methods to Mildly Nonlinear Problems. Ph. D. Diss., Univ of Maryland, College Park, Maryland (1969)

5. Douglas, J., Kellog, R.B., Varga, R.S.: Alternating direction Iteration Methods for n Space Variables. Math. Comp. 17, 279-282 (1963)

6. Guilinger, W.H., Jr.: The Peaceman-Rachford Method for Small Mesh Increments. J. Math. Anal. 11, 261-277 (1965)

7. Habetler, G.J.: Concerning the Implicit Alternating Direction Method. Report KAPL-2040, Knolls Atomic Power Laboratory, Schenectady, New York (1959)

8. More, J.M.: Global Convergence of Newton-Gauss-Seidel Methods. SIAM J. Numer. Anal.8, 325-336 (1971)

9. Ortega, J.M., Rheinboldt, W.C.: Iterative Solution of Nonlinear Equations in Several Variables. New York: Academic Press 1970

10. Ostrowski, A.M.: Über die Determinanten mit überwiegender Hauptdiagonale. Comment. Math. Helv. 10, 69-96 (1937)

11. Pearcy, C.: On Convergence of Alternating Direction Procedures. Numer. Math. 4, 172-176 (1962)

12. Price, H., Varga, R.S.: Recent Numerical Experiments Comparing Successive Overrelaxation Iterative Methods with Implicit Alternating Direction Methods. Report Nr. 91, Gulf Research and Development Co., Pittsburgh, Pennsylvania (1962)

13. Varga, R.S.: Matrix Iterative Analysis. Series in Automatic Computation, Englewood Cliffs., N.J.: Prentice Hall, (1962)

14. Wachspress, E.: Iterative Solution of Elliptic Systems. Englewood Cliffs, N.J.: Prentice Hall 1966

15. Widlund, O.B.: On the Rate of Convergence of an Alternating Direction Implicit Method in a Noncommutative Case. Math. Comp. 20, 500-515 (1966)

16. Young, D.M.: Iterative Solution of Large Linear Systems. New York: Academic Press 1971

Efficient Embedded Runge - Kutta Methods

D. G. Bettis

Texas Institute for Computational Mechanics
The University of Texas at Austin
Austin, Texas 78712
U S A

I. Introduction

For the initial value problem for a system of ordinary differential equations

$$\frac{dx}{dt} = f(t,x) \ , \quad x(t_o) = x_o \ ,$$

the Runge-Kutta algorithm of order p is

$$x_p(t_o + h) = x_o + h \sum_{k=o}^{r_p} C_{p,k} \, f_{p,k} + O(h^{p+1})$$

where

$$f_{p,o} = f(t_o, x_o)$$

$$f_{p,k} = f(t_o + \alpha_{p,k} h, x_o + h \sum_{\lambda=o}^{k-1} \beta_{p,k,\lambda} \, f_{p,\lambda}), \quad k = 1,2,\ldots,r_p \ .$$

The coefficients α, β and C are selected so that the approximate solution x_p is identical to that given by a Taylor sum

$$x_p(t_o + h) = \sum_{\nu=o}^{p} h^\nu \, X_\nu \ ,$$

where $X_o = x_o$, $X_1 = f(t_o, x_o)$, and for $\nu > 1$

$$X_\nu = \left. \frac{d^{(\nu-1)} f}{dt^{(\nu-1)}} \right|_{t_o, x_o} \ .$$

The difference between two solutions x_p and x_{p+1}, yields an estimate of the term $h^{p+1} X_{p+1}$. This difference, ER_{p+1} is therefore an estimate of either i) the leading term of the truncation error of the solution x_p , or ii) of the last term of the representative Taylor sum of the solution x_{p+1}. The value $|ER_{p+1}|$ has been used traditionally to compute a new value for the stepsize, h. Additionally, this value may be used to determine which value of p should be used, provided that the coefficients α, β and C are available for a sequence of values of p. If these coefficients are available, a strategy may be adapted for a variable-step, variable-order Runge-Kutta procedure. Such an algorithm will be competative with other methods only if it is efficient, that is, for a specified accuracy tolerance, TOL, the global error should be proportional to TOL at a minimum of computation time. The purpose of this presentation is to consider the selection of the coefficients α, β and C that lead to efficient Runge-Kutta algorithms of this type, where $p = 1,2,\ldots,p_\nu$. For this study, coefficients will be presentated for p_ν equal to six.

I.a) Number of Stages

The expense of a single step may be measured by the amount of computation time during the step, the total number of operations required for both the Runge-Kutta algorithm and for the function evaluations, or the number of function evaluations only. No matter what the viewpoint may be, the number of evaluations per step, or stages, $s_p = r_p + 1$, should be as few as possible. It may be, however, that the efficiency of the method can be enhanced by increasing the number of stages so that it has more favorable characteristics regarding truncation errors and stability, as will be discussed.

For a Runge-Kutta method that has embedded in it two or more solutions of different order, it is essential that as many of the function evaluations as possible be in common for the different solutions. That is, as far as possible, the $f_{p,k}$ should be identical for the various values of p.

Considering two embedded solutions, x_p, x_{p+1}, the new stepsize determined from ER_{p+1} may be varied with either solution, so that x_o for the new step may represent x_p or x_{p+1}. For the first situation, the last function evaluation may be used for the first function evaluation of the subsequent step if

$$r_{p+1} > r_p,\ \alpha_{p+1,r_{p+1}} = 1\ ,\ \text{and}\ C_{p,k} = \beta_{p+1,r_{p+1},k},\ k = 0,1,\ldots,r_p.$$

Similary, if the higher order solution is being used, and if $r_p > r_{p+1}$, $\alpha_{p,r_p} = 1$, and $C_{p+1,k} = \beta_{p,r_p,k}$, $k = 0,1,\ldots,r_{p+1}$,

then, again, the last and first function evaluations are identical for two successive steps. In either case, this technique reduces the number of function evaluations by one, after the initial step.

I.b) Local Truncation Error Terms

The difference between the Runge-Kutta solution x_p and the Taylor series approximation of the solution may be expressed as

$$h^{p+1}\sum_{i=1}^{\lambda_{p+1}} T_{p,p+1,i}\,E_{p+1,i} + h^{p+2}\sum_{i=1}^{\lambda_{p+2}} T_{p,p+2,i}\,E_{p+2,i} + \ldots ,$$

where the T terms depend upon the coefficients α, β and C associated with the particular Runge-Kutta solution, where the $E_{q,i}$ terms are combinations of the problem-dependent partial derivatives of the total derivative of f of order q, and where λ_q is the number of these derivatives for an autonomous system. The T terms are presented in the report by Bettis and Horn [1976], and a FORTRAN subroutine is provided for their computation up to T_{10}.

The coefficients α, β and C should be selected so that the T terms are small, thereby reducing the magnitude of the leading term of the truncation error of the approximation x_p. The reliability of the error estimate may be affected in the process, however. For example, consider two solutions x_p and x_{p+1}. The estimate ER_{p+1} becomes

$$h^{p+1}\sum_{i=1}^{\lambda_{p+1}} T_{p,p+1,i}\,E_{p+1,i} + h^{p+2}\sum_{i=i}^{\lambda_{p+2}} \delta T_{p+2}\,E_{p+2,i} + \ldots ,$$

where $\delta T_{p+2} = T_{p,p+2,i} - T_{p+1,p+2,i}$. As the $T_{p,p+1,i}$ terms approach zero, as is possible (depending upon the selection of the coefficients and the structure of the solution of the equations of condition), the estimate of ER_{p+1} depends upon the coefficients associated with terms h^{p+2} and higher. Specifically, for ER_{p+1} to estimate the error reliably, the δT_{p+2} terms should approximate the $T_{p,p+2,i}$ terms if the lower order solution is being used. Thus, as a consequence of minimizing the terms $T_{p,p+1,i}$, the estimate of the leading truncation error term by δT_{p+2} may become misleading [Bettis, 1976]. For $ER_{p+1,i}$ to be a reliable estimate of the last term of the Taylor sum for the solution x_{p+1}, none of the terms $T_{p,p+1,i}$ should be zero, because, if they are, the corresponding $E_{p+1,i}$ term will not be accounted for in the estimate, although it may be large, and possibly the dominant partial derivative contributing to the error. Thus, when the higher order solution is being utilized, the coefficients of the lower order solution should be selected so that the $T_{p,p+1,i}$ terms do not vanish.

Since ER_{p+1} will be a reliable estimate of the error so long as the $T_{p,p+1,i}$ terms do not vanish, it will be advantageous to select the coefficients so that components of the leading error term of the higher order solution are small, so that by using the higher order solution x_{p+1}, the effective accuracy will become of order $p+2$ as the error term becomes negligible. For embedded solutions of order $1,2,\ldots,p+1$, the coefficients should be selected so that, beginning with the higher order solution, the solutions of alternating orders have small truncation terms.

I.c) Stability

For the equation $\frac{dx}{dt} = \lambda x, \lambda$ complex, the solution x_p becomes

$$x_p(t_o + h) = x_o R + O(h^{p+1})$$

where

$$R = \sum_{i=o}^{p} \frac{(\lambda h)^i}{i!} + \sum_{i=p+1}^{s_p} \gamma_i (\lambda h)^i ,$$

and where the γ_i depend upon the coefficients α, β, C. If x_o has an error ε_o at t_o, say $z_o = x_o + \varepsilon_o$, then the difference between $x_p(t_o + h)$ and $z_p(t_o + h)$ becomes

$$\varepsilon_{t_o+h} = \varepsilon_o R .$$

If $|R| < 1$, the solution x_p is defined to be absolutely stable. For values of h, λ and γ_i such that the solution x_p is absolutely stable, an error in x_o will not be amplified into the next step. For a system of equations, λ becomes the eigenvalues of the constant Jacobian matrix [HENRICI, 1962]. The coefficients of the Runge-Kutta method should be selected so that the stability region, $|R| = 1$ in the complex plane λh is large [STETTER. 1973].

The coefficient dependent γ_i are expressed as

$$\gamma_i = \sum_{i_{i-1}=i-1}^{s_p} c_{i_{i-1}} \sum_{i_{i-2}=i-2}^{i_{i-1}-1} \beta_{i_{i-1}, i_{i-2}} \cdots \sum_{i_2=2}^{i_3-1} \beta_{i_3, i_2} \sum_{i_1=1}^{i_2-1} \beta_{i_2, i_1} \alpha_{i_1} .$$

Each γ_i is related to one of the λ_{p+i} truncation error terms, k, $T_{p+i,k} = \gamma_i - 1/(p+1)!$. Unfortunately, the values of γ_i are not equal to $1/(p+1)!$ for large stability regions. Thus, there is a compromise necessary between minimizing the truncation error terms and maximizing the stability region.

Explicit Runge-Kutta methods are generally not recommended for problems that are characterized by stability difficulties, i.e. stiff equations. However, for large systems of mildly stiff equations, explicit Runge-Kutta methods of low order with $s_p > p$ may prove to be efficient if the absolute stability regions are large, because i) as s_p increases with respect to p, the stability regions can increase, depending upon the values of the γ_i terms, and ii) the method, since it is explicit, does not require an iterative procedure. It must be emphasized that as the larger stability regions are produced because of an increasing number of function evaluations the efficiency of the Runge-Kutta method is diminished.

It is noted that the concept of absolute stability is based on linear differential equations, and that the stability behavior of a method for a nonlinear equation will not necessarily be applicable. In designing a Runge-Kutta method, the coefficents should be selected so that, at least for the linear differential equation, the method possesses absolute stability.

If the type of problems to be solved are of the form of a perturbed harmonic oscillator, the method should be absolutely stable for the pure harmonic oscillator

$$\frac{d^2x}{dt^2} + \omega^2 x = 0,$$

or, as a system of first order equations,

$$\frac{dv}{dt} = +\lambda^2 x \ , \ \frac{dx}{dt} = v \ ,$$

where $\lambda = i\omega$. Since λ is imaginary, the stability region in the vicinity of the imaginary axis should be large. Often, the coefficients of a method have been selected without consideration of the absolute stability region with the result that the region does not cross the imaginary axis, but approaches it asymptotically. These formulas are characterized by their poor performance for the test problem of a harmonic oscillator.

Similary, for nonlinear problems, that have linearized solutions that are characterized by real values of λ , the stability regions should have a large interval along the negative real axis.

I.d) Numerical Considerations

The values of the coefficients β and C should be positive and of the same order of magnitude in order to reduce errors due to round off. Because of the structure of the solution for the Runge-Kutta method, this is usually not possible. As a compromise, the C coefficients should at least be positive, and the β and C coefficients should not differ by more than an order of magnitude.

Since stepsize and order selection are based upon the assumption that the derivatives $E_{p+1,i}$ for a solution x_p are constant throughout a step, values of α should be $0 \le \alpha \le 1$, to minimize the effects of rapidly changing derivatives.

II. An Embedded Method of Order One-Six

II.a) Truncation Error

For the basis of a new embedded method of orders one through six, the method due to Fehlberg[1969] for a fifth and sixth order pair will be adapted. The number of stages for the new embedded method and the minimum number of stages for a method of order p are given below:

p	1	2	3	4	5	6
$s_p(MIN)$	1	2	3	4	6	7
s_p	1	3	4	5	6	7

If the method is to be used as a sixth order method only, without a companion method(s) of lower order(s), then only seven function evaluations are required. However, if solutions of order one through six are desired, two additional evaluations are necessary, i.e. a total of nine evaluations are required for embedded methods of order one through six. When using only a fifth and sixth order solution, only eight function evaluations need to be calculated since $f_{4,4}$ is necessary only for a solution of order four.

The coefficients for the solution of orders two through six depend upon the two parameters $\alpha_{p,1}$ and $\alpha_{p,2}$, $p = 2,3,4,5,6$. The values $\alpha_{p,1} = 0.1423$ and $\alpha_{p,2} = 0.2457$ yield a minimum value of $G_{6,7} = 0.6103E-03$, where

$$G^2_{p,q} = \sum_{i=1}^{\lambda_q} T^2_{p,q,i} .$$

For $\alpha_{p,1} = 1/16$, $\alpha_{p,2} = 4/15$, values selected by Fehlberg, $G_{6,7}$ is $0.1192E-02$. With $\alpha_{p,1} = 4/25$ and $\alpha_{p,2} = 1/4$, $G_{6,7}$ becomes $0.6187E-03$ and, for these values of the two parameters, the coefficients for the solutions of order $p = 1,\dots,6$ are presented in Tables I and II. The value of the parameters should be determined that minimize $G_{p,p+1}$ for additional values of p so that the solution of orders other than six also have small truncation error terms.

II.b) Stability

For the solution of order six with s_p equal to seven,

$$R = \sum_{i=0}^{6} \frac{(\lambda h)^i}{i!} + \gamma_7 (\lambda h)^7 ,$$

where

$$\gamma_7 = c_{6,7}\, \beta_{6,7,6}\, \beta_{6,6,5}\, \beta_{6,5,4}\, \beta_{6,4,3}\, \beta_{6,3,2}\, \beta_{6,2,1}\, \alpha_{6,1} .$$

For the particular solution considered, the expression for γ_7 can be reduced to

$$\frac{\alpha_2(1 - 3\alpha_2)}{720(15\alpha_2^2 - 10\alpha_2 + 2)} ,$$

with $\alpha_2 = \alpha_{p,2}$. Thus, the region of absolute stability depends upon the single parameters α_2. Defining $K = 7!\gamma_7$, the largest interval of absolute stability on the real axis $[-6.511, 0]$ corresponds to a value of $K = 0.547$, or to the values of $\alpha_2 = 0.1234,\ 0.3036$. Unfortunately, for either value of α_2, one ore more of the other coefficients become very large. Also, for these values of α_2, the truncation error term is not as small as possible. In fact, only when α_2 is $1/4$ or $1/9$, is γ_7 identical to $1/7!$, but in this case, the interval of absolute stability on the real axis is $[-3.954, 0]$. Table III presents values of α_2 and their corresponding values of K and R^*, the value of the interval of absolute stability on the negative real axis.

If $\alpha_{p,1} = 1/5$ and $\alpha_{p,2} = 91/300$, then $R^* = -6.439$ and $G_{6,7} = 0.1944E-02$. The selection made by Fehlberg (where absolute stability and the minimization of $G_{6,7}$ were not considered) was $\alpha_{p,1} = 1/6$ and $\alpha_{p,2} = 4/15$, which corresponds to $R^* = -4.065$ and to $G_{6,7} = 0.1192E-02$. Thus, for a set of coefficients with a large stability interval on the real axis, and a reasonable small value of G, the set of coefficients with $\alpha_{p,1} = 1/5$ and $\alpha_{p,2} = 91/300$ may be chosen. Also, the free parameters should be selected so that maximum absolute stability regions are obtained for the solutions of order other than six, if possible.

II.c) Quadrature

If the pair of solutions x_5 and x_6 are being used to solve a quadrature problem, then the difference between the two solutions is identically zero. This difficulty arises because both the x_5 and x_6 solutions are of order six for the special case of a quadrature. This occurance is easily detected [Shampine] and, for the coefficients presented in Tables I and II, the following relation will provide an estimate of the error, of order six, for a quadrature (ER_6 = quad est)

$$\text{quad est} = h\left[\frac{33}{560} f_{6,0} - \frac{859375}{2522016} f_{6,1} + \frac{2816}{7695} f_{6,2} - \frac{26411}{168480} f_{6,3} + \frac{649539}{7209930} f_{6,4} - \frac{11}{630} f_{6,5}\right].$$

<u>References</u>

[1] Bettis, D.G.: Embedded Runge-Kutta Methods of Order four and five, Numer. Math. (in press)

[2] Bettis, D.G. and Horn, K.: Computation of Truncation Error Terms for Runge-Kutta Methods.
Texas Institute for Computational Mechanics Report Series, The University of Texas at Austin, 1976.

[3] Fehlberg, E.: Klassische Runge-Kutta-Formen fünfter und siebenter Ordnung mit Schrittweiten-Kontrolle,
Computing 4 (1969), 93 - 106.

[4] Shampine, L.F.: Quadrature and Runge-Kutta Formulas,
Appl. Math. Computing 2 (1976), 161 - 171.

[5] Stetter, H.J.: Analysis of Discretization Methods for Ordinary Differential Equations,
Springer-Verlag 1973.

<u>Acknowledgements:</u>

This work has been supported by the Division of Mathematical and Computer Sciences, National Science Foundation, Grant DCR 75-17309.

p	$k \backslash \lambda$	$\alpha_{p,k}$	$\beta_{p,k,\lambda}$: 0	1	2	3	4	5
$1 \to 6$	0	0	0					
$2 \to 6$	1	$\frac{4}{25}$	$\frac{4}{25}$					
$2 \to 6$	2	$\frac{1}{4}$	$\frac{7}{128}$	$\frac{25}{128}$				
$3 \to 6$	3	$\frac{4}{7}$	$\frac{170}{343}$	$-\frac{550}{343}$	$\frac{576}{343}$			
4	4	$\frac{7}{9}$	$\frac{203}{729}$	$-\frac{12425}{23328}$	$\frac{4123}{6561}$	$\frac{84721}{209952}$		
5, 6	4	$\frac{7}{9}$	$-\frac{3773}{17496}$	$\frac{1925}{1458}$	$-\frac{17024}{19683}$	$\frac{84721}{157464}$		
5	5	1	$\frac{1165}{4424}$	$-\frac{275}{158}$	$\frac{3808}{1501}$	$-\frac{7595}{8216}$	$\frac{118098}{136591}$	
6	5	0	$-\frac{81}{1106}$	0	$\frac{288}{1501}$	$-\frac{441}{2054}$	$\frac{13122}{136591}$	
6	6	1	$-\frac{2935}{4424}$	$-\frac{275}{158}$	$\frac{3520}{1501}$	$-\frac{5831}{8216}$	$\frac{104976}{136591}$	1

Table I . α and β coefficients

$k\backslash p$	$c_{p,k}$					
	1	2	3	4	5	6
0	1	-1	$\frac{11}{24}$	$\frac{5}{168}$	$\frac{43}{560}$	$\frac{11}{3024}$
1		0	0	0	0	
2		2	$-\frac{16}{27}$	$\frac{272}{513}$	$\frac{2816}{7695}$	
3			$\frac{245}{216}$	$-\frac{343}{2808}$	$\frac{16807}{84240}$	
4				$\frac{972}{1729}$	$\frac{19683}{69160}$	
5					$\frac{79}{1080}$	
6						$\frac{79}{1080}$

Table II. C coefficients

K	R^*	α_2
.5437	-6.511	.1234, .3036
.5511	-6.415	.1242, .3033
.5553	-6.310	.1250, .3030
.5602	-6.192	.1258, .3027
.5657	-6.059	.1269, .3024
.5723	-5.9o4	.1281, .3019
.5804	-5.720	.1297, .3014 .

Table III. α_2 *vs.* K and R^*

COLLOCATION AND ITERATED DEFECT CORRECTION

Reinhard Frank, Christoph W. Ueberhuber
Inst. f. Num. Math., Technical University of Vienna

1. INTRODUCTION

In this paper an equivalence between solutions of collocation methods and fixed points of Iterated Defect Correction (IDeC) methods is proved. Therefore the IDeC-methods can be regarded as efficient schemes for solving collocation equations. Attention is restricted to the application of the IDeC to ordinary differential equations (initial value problems and two point boundary value problems). Extension to other types of operator equations (e.g. partial differential equations, integral equations,...) is straightforward.

In Section 2 special variants of collocation methods, which are of importance in connection with the IDeC are discussed. The basic ideas behind the IDeC are presented in Section 3. The equivalence between collocation schemes and the fixed points of the IDeC-methods is established in Section 4.

2. COLLOCATION METHODS

2.1. Collocation methods for two point boundary value problems

We consider problems of the form

(2.1a) $$y' = f(t,y), \quad t \in [a,b]$$

(2.1b) $$g(y(a),y(b)) = 0$$

where y,f and g are vector-valued functions of dimension n with f and g sufficiently smooth. A number of papers about collocation methods applied to (2.1) have appeared recently in the literature on the numerical solution of BVPs for ODEs (e.g. de Boor, Swartz [3], Russel, Shampine [9], Weiss [11]). From the class of collocation schemes, we consider the following special type (cf. Weiss [11]):

The collocation solution is a continuous piecewise polynomial which satisfies (2.1a) at given (collocation) points.

We now introduce the notation to be used below. The grid is given by

$$(2.2)\qquad \begin{aligned} &a = t_0 < \ldots < t_I = b \\ &H_i := t_{i+1} - t_i, \qquad i = 0(1)I-1. \end{aligned}$$

We consider the space of continuous piecewise polynomial functions P(t) [vector-valued of dimension n] defined by

$$(2.3)\qquad \begin{aligned} &P(t) := P_i(t), \qquad t \in [t_i, t_{i+1}], \qquad i = 0(1)I-1 \\ &P_i(t_{i+1}) = P_{i+1}(t_{i+1}), \qquad i = 0(1)I-2 \end{aligned}$$

where *all* polynomials P_i are of *degree m*. On (2.2) we construct the subgrid

$$(2.4)\qquad t_{i,k} := t_i + \xi_k H_i, \qquad i = 0(1)I-1, \qquad k = 1(1)m$$

with collocation nodes

$$(2.5)\qquad 0 \le \xi_1 < \ldots < \xi_m \le 1$$

(important special cases satisfying (2.5) are the Gauss-Legendre, the Lobatto and the Radau points). The collocation equations become

$$(2.6)\qquad P_i'(t_{i,k}) = f(t_{i,k}, P_i(t_{i,k})), \qquad i = 0(1)I-1, \quad k = 1(1)m.$$

If $\xi_1 = 0$ or $\xi_m = 1$ in (2.5), then $P_i'(t_{i,1})$ or $P_i'(t_{i,m})$ is interpreted as the right derivative or the left derivative, respectively. If $\xi_1 = 0$ *and* $\xi_m = 1$, then two collocation equations (2.6) hold at every gridpoint $t_i = t_{i,1} = t_{i-1,m}$. Together with the boundary condition

$$(2.7)\qquad g(P_0(a), P_{I-1}(b)) = 0$$

and the continuity conditions

$$(2.8)\qquad P_i(t_{i+1}) = P_{i+1}(t_{i+1}), \qquad i = 0(1)I-2,$$

the collocation conditions yield $n \cdot I \cdot (m+1)$ equations for the $n \cdot I \cdot (m+1)$ unknown coefficients of P.

2.2. Collocation methods for initial value problems

The method of Section 2.1 can be interpreted as a method for solving IVPs, if in (2.1) the boundary condition is replaced by

$$(2.9)\qquad g(y(a), y(b)) = y(a) - y_a = 0.$$

In this situation, it is possible to solve the equations (2.6) block by block (where one block contains the equations for the interval $[t_i, t_{i+1}]$).

Collocation methods can only be justified as efficient computational strategies for IVPs when the equations are stiff. For such equations these methods have the advantage of good stability properties combined with high order accuracy (cf. e.g. Wright [12], Axelsson [1], Ehle [4], Chipman [2]).

3. ITERATED DEFECT CORRECTION

The Iterated Defect Correction (IDeC) consists essentially in an iterative improvement of a given numerical solution (obtained from some finite difference method). In this section we describe methods derived from this concept for which an equivalence with collocation methods will be established below.

3.1. IDeC-methods for two point boundary value problems

In order to obtain an initial approximation to (2.1) by some finite difference method, we introduce a grid which is a refinement of grid (2.2):

$$(3.1)\qquad \left.\begin{array}{l} t_i = s_{i,o} < \ldots < s_{i,K} = t_{i+1} \\ h_{i,k} := s_{i,k+1} - s_{i,k}\,, \quad k = O(1)K-1 \end{array}\right\} \quad i = O(1)I-1$$

Note: The points t_i have two different names in the notation of (3.1):

$$(3.2)\qquad t_i = s_{i,o} = s_{i-1,K}\,, \quad i = 1(1)I-1$$

On the grid (3.1), we next consider three well-known finite difference methods.

For $i = O(1)I-1, \quad k = O(1)K-1$:

$$(3.3)\qquad \begin{cases} (\eta_{i,k+1} - \eta_{i,k})/h_{i,k} = f(s_{i,k} + h_{i,k}/2, (\eta_{i,k} + \eta_{i,k+1})/2) \\ g(\eta_{o,o}, \eta_{I-1,K}) = 0 \end{cases}$$

$$(3.4)\qquad \begin{cases} (\eta_{i,k+1} - \eta_{i,k})/h_{i,k} = f(s_{i,k}, \eta_{i,k}) \\ g(\eta_{o,o}, \eta_{I-1,K}) = 0 \end{cases}$$

$$(3.5)\qquad \begin{cases} (\eta_{i,k+1}-\eta_{i,k})/h_{i,k} = f(s_{i,k+1},\eta_{i,k+1}) \\ g(\eta_{o,o},\eta_{I-1,K}) = 0. \end{cases}$$

Because of (3.2), we require (for all three methods) that

$$(3.6)\qquad \eta_{i,o} = \eta_{i-1,K}\ , \quad i = 1(1)I-1$$

which ensures that the number of equations and the number of unknowns are indentical.

These methods are usually defined on grids with only one index. However, the more involved notation of (3.1) will turn out to be essential for the analysis in the next section.

The solution of a given finite difference scheme ((3.3), (3.4) or (3.5)) is denoted by

$$\eta^{o} := \left(\eta^{o}_{o,o},\ldots,\eta^{o}_{o,K},\eta^{o}_{1,o},\eta^{o}_{1,1},\ldots,\eta^{o}_{I-1,K}\right).$$

Interpolation of η^{o} by a piecewise polynomial function P^{o}

$$(3.7)\qquad \left.\begin{aligned} P^{o}(t) &:= P^{o}_{i}(t), & t\in[t_i,t_{i+1}] \\ P^{o}_{i}(s_{i,k}) &= \eta^{o}_{i,k}\ , & k = 0(1)K \end{aligned}\right\}\ i = 0(1)I-1$$

yields the defect

$$(3.8)\qquad \begin{aligned} d^{o}_{i}(t) &:= (P^{o}_{i})'(t) - f(t,P^{o}_{i}(t)), \\ &\qquad t\in[t_i,t_{i+1}],\quad i = 0(1)I-1. \end{aligned}$$

By adding the defect d^{o}_{i} to the righthand side of the original problem (2.1), we obtain a new BVP of a slightly more general type:
From the set of continuous piecewise functions

$$(3.9)\qquad \begin{aligned} \{y\,|\,y(t) &:= y_i(t),\ t\in[t_i,t_{i+1}], \quad i = 0(1)I-1; \\ y_i(t_{i+1}) &= y_{i+1}(t_{i+1}),\quad i = 0(1)I-2\} \end{aligned}$$

(where sufficiently high derivatives of the functions y_i exist), we determine that function which satisfies the following relations

$$(3.10)\qquad \begin{aligned} & y_i' = f(t,y_i) + d^{o}_{i}(t),\quad t\in[t_i,t_{i+1}],\quad i = 0(1)I-1 \\ & g(y_0(a),y_{I-1}(b)) = 0 \end{aligned}$$

The *exact* solution of this "piecewise" BVP is P^{o} (cf. (3.8)). Despite our knowledge of the exact solution, we solve the new BVP (3.10) in the same way as the original BVP (2.1), i.e. the same finite difference scheme [(3.3), (3.4) or (3.5)] which was used to obtain η^{o}, is now applied to (3.10). This yields

$$\pi^{o} = \left(\pi^{o}_{o,o},\ldots,\pi^{o}_{o,K},\ \pi^{o}_{1,o},\ \pi^{o}_{1,1},\ldots,\pi^{o}_{I-1,K}\right).$$

We can now use the *known* global discretization errors $\pi^{o}_{i,k} - P^{o}_{i}(s_{i,k})$ of (3.10) as estimates for the unknown global discretization errors $\eta^{o}_{i,k} - y(s_{i,k})$ of (2.1). The original idea of estimating the global discretization error in this way is due to Zadunaisky [13].

If we replace the unknown error term in the identity

$$(3.11)\qquad y(s_{i,k}) = \eta^{o}_{i,k} - \left(\eta^{o}_{i,k} - y(s_{i,k})\right).$$

by our estimate, we obtain the following formula for the improvement of our first solution η^{o}:

$$(3.12)\qquad \eta^{1}_{i,k} := \eta^{o}_{i,k} - \left(\pi^{o}_{i,k} - P^{o}_{i}(s_{i,k})\right)$$

The whole procedure may be used iteratively,

$$(3.13)\qquad \eta^{j+1}_{i,k} := \eta^{o}_{i,k} - \left(\pi^{j}_{i,k} - P^{j}_{i}(s_{i,k})\right),\qquad j = 1,2,\ldots$$

where P^{j} denotes the polynomial which interpolates η^{j} (analog to (3.7)).

The above iterative strategy is called the *Iterated Defect Correction (IDeC)*, and the different methods which can be constructed using this concept are called *IDeC-methods*. More details about the IDeC and IDeC-methods are available (see, for example Stetter [8] or Frank, Ueberhuber [6]).

The IDeC-methods described above use estimates of the *global* discretization error. We will now discuss other IDeC-methods which use estimates of the *local* discretization errors. As in the "global case", we start with η^{o} (solution of (3.3), (3.4) or (3.5)), interpolate η^{o} by P^{o} and construct the new BVP (3.10) the exact solution of which is P^{o}. Therefore, the exact local discretization error associated with problem (3.10) can be evaluated: e.g., for the box-scheme (3.3), we obtain

$$
\begin{aligned}
l^{o}_{i,k} :=\ & \left(P^{o}_{i}(s_{i,k+1}) - P^{o}_{i}(s_{i,k})\right)/h_{i,k} - \\
& - f\,(s_{i,k}+h_{i,k}/2, [P^{o}_{i}(s_{i,k}) + P^{o}_{i}(s_{i,k+1})]/2) - \\
& - \left(P^{o}_{i}\right)'(s_{i,k}+h_{i,k}/2) + \\
& + f\,(s_{i,k}+h_{i,k}/2,\ P^{o}_{i}(s_{i,k}+h_{i,k}/2)) = \\
=\ & - \left(P^{o}_{i}\right)'(s_{i,k}+h_{i,k}/2) + f(s_{i,k}+h_{i,k}/2,\ P^{o}_{i}(s_{i,k}+h_{i,k}/2)) = \\
=\ & - d^{o}_{i}(s_{i,k}+h_{i,k}/2)
\end{aligned}
\tag{3.14}
$$

as an estimate for the unknown local discretization error of (2.1)

$$
\begin{aligned}
l_{i,k} :=\ & (y(s_{i,k+1}) - y(s_{i,k}))/h_{i,k} - \\
& - f\,(s_{i,k}+h_{i,k}/2,\ [y(s_{i,k}) + y(s_{i,k+1})]/2).
\end{aligned}
\tag{3.15}
$$

To obtain the improved approximation η^{1}, it is necessary to solve

$$
\begin{aligned}
& \left(\eta^{1}_{i,k+1} - \eta^{1}_{i,k}\right)/h_{i,k} - f(s_{i,k}+h_{i,k}/2, [\eta^{1}_{i,k} + \eta^{1}_{i,k+1}]/2) = \\
& \qquad = l^{o}_{i,k} = - d^{o}_{i}(s_{i,k}+h_{i,k}/2) \\
& g\left(\eta^{1}_{o,o},\ \eta^{1}_{I-1,K}\right) = 0, \quad \eta^{1}_{i,K} = \eta^{1}_{i+1,o}.
\end{aligned}
\tag{3.16}
$$

This procedure may again be used iteratively, yielding η^{2}, η^{3},... . The error estimate used in obtaining η^{2} is

$$
\begin{aligned}
l^{1}_{i,k} =\ & \left(P^{1}_{i}(s_{i,k+1}) - P^{1}_{i}(s_{i,k})\right)/h_{i,k} - \\
& - f\,(s_{i,k}+h_{i,k}/2,\ [P^{1}_{i}(s_{i,k}) + P^{1}_{i}(s_{i,k+1})]/2) - \\
& - \left(P^{1}_{i}\right)'(s_{i,k}+h_{i,k}/2) + \\
& + f\,(s_{i,k}+h_{i,k}/2,\ P^{1}_{i}(s_{i,k}+h_{i,k}/2)) = \\
=\ & \left(\eta^{1}_{i,k+1} - \eta^{1}_{i,k}\right)/h_{i,k} - \\
& - f\,(s_{i,k}+h_{i,k}/2,\ [\eta^{1}_{i,k} + \eta^{1}_{i,k+1}]/2) - \\
& - \left(P^{1}_{i}\right)'(s_{i,k}+h_{i,k}/2) + \\
& + f\,(s_{i,k}+h_{i,k}/2,\ P^{1}_{i}(s_{i,k}+h_{i,k}/2)) = \\
=\ & - d^{o}_{i}(s_{i,k}+h_{i,k}/2) - d^{1}_{i}(s_{i,k}+h_{i,k}/2).
\end{aligned}
\tag{3.17}
$$

The general formula for the error estimate is

$$(3.18)\qquad l^j_{i,k} = -d^o_i(s_{i,k}+h_{i,k}/2) - \ldots - d^j_i(s_{i,k}+h_{i,k}/2).$$

Note This method is a special case of the difference correction of Fox and Pereyra (cf. e.g. Pereyra [8]). Another "local version" of the IDeC which is more similar to Pereyra's approach has been discussed by Frank, Hertling, Ueberhuber [7].

We now establish an equivalence result for both the above variants.

THEOREM 3.1. *Consider the following general formulation for a linear BVP*

$$(3.19)\qquad \begin{aligned} y' &= Ay \\ By(a) + Cy(b) &= e. \end{aligned}$$

Then the approximations $\eta^1, \eta^2, \ldots$ given by both the above mentioned variants of the IDeC are identical.

PROOF: For the linear equations (3.19), the schemes (3.3), (3.4) and (3.5) [together with (3.6)] may be written as

$$(3.20)\qquad D\eta^o = \begin{pmatrix} 0 \\ e \end{pmatrix}.$$

For example, in the case of the box-scheme, (3.20) becomes

$$(3.21)\qquad \begin{aligned} \left(\eta^o_{i,k+1} - \eta^o_{i,k}\right)/h_{i,k} - (1/2)A\left(\eta^o_{i,k} + \eta^o_{i,k+1}\right) &= 0 \\ \eta^o_{i,K} - \eta^o_{i+1,o} &= 0 \\ B\eta^o_{o,o} + C\eta^o_{I-1,K} &= e. \end{aligned}$$

We use induction. For j = 1, we obtain

(i) "local variant":

$$D\eta^1 = \begin{pmatrix} 0 \\ e \end{pmatrix} - \begin{pmatrix} d^o \\ 0 \end{pmatrix} \qquad \text{(cf. (3.14), (3.16))}$$

i.e.

$$\eta^1 = D^{-1}\begin{pmatrix} 0 \\ e \end{pmatrix} - D^{-1}\begin{pmatrix} d^o \\ 0 \end{pmatrix}$$

Note 1 For example, in the case of the box-scheme, d^o becomes

$$d^o := \left(d^o_o(s_{o,o}+h_{o,o}/2),\ldots,d^o_{I-1}(s_{I-1,K-1}+h_{I-1,K-1}/2)\right).$$

Note 2 If (3.19) has a unique solution, then it is well known that (3.20) has also a unique solution for sufficiently fine grids, i.e. D^{-1} is defined.

(ii) "global variant":

$$D\pi^o = \begin{pmatrix} 0 \\ e \end{pmatrix} + \begin{pmatrix} d^o \\ 0 \end{pmatrix} \qquad \text{(cf. (3.10))}$$

$$\pi^0 = D^{-1}\begin{pmatrix}0\\e\end{pmatrix} + D^{-1}\begin{pmatrix}d^0\\0\end{pmatrix}$$

$$\eta^1 = \eta^0 - (\pi^0 - \eta^0) = \qquad \text{(cf. (3.12))}$$

$$= 2\eta^0 - \pi^0 =$$

$$= 2D^{-1}\begin{pmatrix}0\\e\end{pmatrix} - \left[D^{-1}\begin{pmatrix}0\\e\end{pmatrix} + D^{-1}\begin{pmatrix}d^0\\0\end{pmatrix}\right] =$$

$$= D^{-1}\begin{pmatrix}0\\e\end{pmatrix} - D^{-1}\begin{pmatrix}d^0\\0\end{pmatrix}$$

which establishes the identity for the case $j = 1$. Let us now assume, for $j = r$, that

$$\eta^r = D^{-1}\left[\begin{pmatrix}0\\e\end{pmatrix} - \begin{pmatrix}d^0\\0\end{pmatrix} - \dots - \begin{pmatrix}d^{r-1}\\0\end{pmatrix}\right]$$

is valid for both variants of the IDeC. Then for $j = r+1$, we obtain

(i) <u>"local variant"</u>:

$$\eta^{r+1} = D^{-1}\left[\begin{pmatrix}0\\e\end{pmatrix} - \begin{pmatrix}d^0\\0\end{pmatrix} - \dots - \begin{pmatrix}d^r\\0\end{pmatrix}\right] \qquad \text{(cf. (3.18))}.$$

(ii) <u>"global variant"</u>:

$$D\pi^r = \begin{pmatrix}0\\e\end{pmatrix} + \begin{pmatrix}d^r\\0\end{pmatrix}$$

$$\pi^r = D^{-1}\begin{pmatrix}0\\e\end{pmatrix} + D^{-1}\begin{pmatrix}d^r\\0\end{pmatrix}$$

$$\eta^{r+1} = \eta^0 - (\pi^r - \eta^r) =$$

$$= D^{-1}\begin{pmatrix}0\\e\end{pmatrix} - D^{-1}\begin{pmatrix}0\\e\end{pmatrix} - D^{-1}\begin{pmatrix}d^r\\0\end{pmatrix} +$$

$$+ D^{-1}\begin{pmatrix}0\\e\end{pmatrix} - D^{-1}\begin{pmatrix}d^0\\0\end{pmatrix} - \dots - D^{-1}\begin{pmatrix}d^{r-1}\\0\end{pmatrix} =$$

$$= D^{-1}\left[\begin{pmatrix}0\\e\end{pmatrix} - \begin{pmatrix}d^0\\0\end{pmatrix} - \dots - \begin{pmatrix}d^r\\0\end{pmatrix}\right]$$

which proves the assertion. ■

<u>3.2. IDeC-methods for initial value problems</u>

If, in Section 3.1, the special boundary condition

$$(3.22) \qquad g(y(a), y(b)) = y(a) - y_a = 0$$

is used, certain IDeC-methods for IVPs are immediately defined. The schemes (3.3), (3.4) and (3.5) are now the implicit midpoint rule, the explicit Euler method and the implicit Euler method. Other IDeC-methods for IVPs are obtained when more general RK-methods are used. This is discussed in Frank, Ueberhuber [5] where the following asymptotic result is proved for equidistant grids ($H_i \equiv H = (b-a)/I$, $h_{i,k} \equiv h = H/K$):

THEOREM 3.2. *If a RK-scheme of order* $p (\leq K)$ *is used, and if* f *satisfies suitable smoothness conditions, then*

$$\eta^j_{i,k} - y(s_{i,k}) = O\left(h^{\min(p(j+1),K)}\right) \quad \text{for } h \to 0. \qquad (3.23)$$ [1]

Note We interpret "$h \to 0$" in the sense of "$I \to \infty$ and K fixed".

We have introduced the IDeC as an iterative scheme, but up to now we have not dicussed how to terminate the process. According to (3.23), a reasonable termination criterion is given by the maximum achievable order K, which is reached for η^J if $K/p \in (J,J+1]$. In section 4, other termination criteria will be discussed.

For BVPs, each IDeC-method consists in computing successively each of the iterates $\eta^1, \eta^2, \ldots$ for the whole interval $[a,b]$. For IVPs, it is of course possible to proceed in a blockwise manner, as is indicated by the following:

$$\begin{bmatrix} \eta^0_{0,0}, & \ldots, & \eta^0_{0,K} \\ \eta^1_{0,0}, & \ldots, & \eta^1_{0,K} \\ \vdots & & \vdots \\ \eta^J_{0,0}, & \ldots, & \eta^J_{0,K} \end{bmatrix}$$
$$\begin{bmatrix} \eta^0_{1,0}, & \ldots, & \eta^0_{1,K} \\ \vdots & & \vdots \\ \eta^J_{1,0}, & \ldots, & \eta^J_{1,K} \end{bmatrix}$$
$$\begin{bmatrix} \vdots & & \vdots \end{bmatrix}$$

The use of such a strategy yields an economy in storage. In Frank, Ueberhuber [6], a more detailed discussion of this procedure may be found.

Just as for BVPs, there exist two possibilities (using either estimates of the *global* or estimates of the *local* discretization error) to construct IDeC-methods for IVPs. Theorem 3.1 may immediately be applied to IVPs, which means that both variants yield identical results for the linear problem.

---

1) K is the degree of the interpolating polynomials P^j_i (cf. (3.7)).

4. RELATIONS BETWEEN COLLOCATION AND IDEC

When IDeC-methods are applied to certain problems, the convergence of the iterates $\eta^1, \eta^2, \ldots$ to a fixed point η^* may be observed. In this section, we will show that those IDeC-methods based on the schemes (3.3), (3.4) or (3.5) (cf. Section 3) have fixed points that coincide with the solutions of collocation schemes (discussed in Section 2).

4.1. Boundary value problems

4.1.1. IDeC-methods based on the box-scheme

We start our discussion by establishing a relationship between the IDeC based on the box-scheme (3.3) and an appropriate collocation scheme. For the IDeC-methods, we assume a grid of the form (3.1) with the same proportional spacing of the subgrid-points $s_{i,k}$ in $[t_i, t_{i+1}]$, i.e. for $k = 1(1)K$

(4.1) $$(s_{o,k} - s_{o,k-1})/H_o = \ldots = (s_{I-1,k} - s_{I-1,k-1})/H_{I-1}.$$

The related collocation is defined on the following grid:

(4.2) $$t_{i,k} := (s_{i,k-1} + s_{i,k})/2, \quad i = 0(1)I-1, \quad k = 1(1)m$$

with

$$m = K.$$

Therefore the corresponding collocation nodes satisfy

(4.3) $$0 < \xi_1 < \ldots < \xi_m < 1.$$

Note A straightforward generalization would consist in dropping the relation (4.1), resulting in a different collocation scheme on every interval $[t_i, t_{i+1}]$.

THEOREM 4.1. *Consider an IDeC-method which uses the box-scheme and a global discretization error estimate. η^* is a fixed point of this IDeC-method iff P^*, defined by*

(4.4) $$P_i^*(s_{i,k}) = \eta^*_{i,k}, \quad i = 0(1)I-1, \quad k = 0(1)m$$

is the solution of the corresponding collocation scheme, i.e.

(4.5) $$d_i^*(t_{i,k}) := (P_i^*)'(t_{i,k}) - f(t_{i,k}, P_i^*(t_{i,k})) = 0,$$
$$i = 0(1)I-1, \quad k = 1(1)m$$

PROOF: By definition, η^* is a fixed point, iff one step of the IDeC-method applied to η^* leads again to η^*, i.e.

$$(4.6)\qquad \eta^*_{i,k} = \overset{\circ}{\eta}_{i,k} - (\pi^*_{i,k} - \eta^*_{i,k}), \qquad i = 0(1)I-1,\quad k = 0(1)m.$$

Fixed points are therefore characterized by

$$(4.7)\qquad \pi^* = \overset{\circ}{\eta}.$$

a) Assume P^* is the solution of the (corresponding) collocation scheme, i.e. (4.5) is satisfied. The equations defining $\overset{\circ}{\eta}$ are

$$(4.8)\qquad \begin{aligned} &(\overset{\circ}{\eta}_{i,k+1} - \overset{\circ}{\eta}_{i,k})/h_{i,k} = f(s_{i,k} + h_{i,k}/2, (\overset{\circ}{\eta}_{i,k} + \overset{\circ}{\eta}_{i,k+1})/2)\\ &\overset{\circ}{\eta}_{i,o} = \overset{\circ}{\eta}_{i-1,m}\\ &g(\overset{\circ}{\eta}_{o,o}, \overset{\circ}{\eta}_{I-1,m}) = 0. \end{aligned}$$

π^* is defined by

$$(4.9)\qquad \begin{aligned} &(\pi^*_{i,k+1} - \pi^*_{i,k})/h_{i,k} = f(s_{i,k} + h_{i,k}/2, (\pi^*_{i,k} + \pi^*_{i,k+1})/2) +\\ &\qquad\qquad + d^*_i(s_{i,k} + h_{i,k}/2)\\ &\pi^*_{i,o} = \pi^*_{i-1,m}\\ &g(\pi^*_{o,o}, \pi^*_{I-1,m}) = 0. \end{aligned}$$

Since $t_{i,k} = s_{i,k-1} + h_{i,k-1}/2$ [cf. (4.2)] and (4.5) is satisfied, the equations (4.8) and (4.9) are identical, i.e. (4.7) is satisfied.

b) Let η^* be a fixed point, i.e. $\overset{\circ}{\eta} = \pi^*$. Subtraction of (4.8) from (4.9) leads immediately to the desired result (4.5). ■

THEOREM 4.2. *Consider an IDeC-method which uses the box-scheme and a local discretization error estimate. η^* is a fixed point of this IDeC-method iff*

$$(4.10)\qquad d^*_i(t_{i,k}) = 0, \qquad i = 0(1)I-1, \qquad k = 1(1)m.$$

PROOF: Let us consider one IDeC-step starting from η^*. The estimate of the local discretization error is

$$(4.11)\qquad \begin{aligned} l^*_{i,k} = {}&(\eta^*_{i,k+1} - \eta^*_{i,k})/h_{i,k} - f(s_{i,k} + h_{i,k}/2, (\eta^*_{i,k} + \eta^*_{i,k+1})/2) -\\ &- d^*_i(s_{i,k} + h_{i,k}/2). \end{aligned}$$

The equations for the next iterate are

$$(4.12)\qquad (\eta^{**}_{i,k+1} - \eta^{**}_{i,k})/h_{i,k} - f(s_{i,k} + h_{i,k}/2, (\eta^{**}_{i,k} + \eta^{**}_{i,k+1})/2) = l^*_{i,k}.$$

a) Suppose (4.10) is satisfied. Then (4.11) and (4.12) imply that $\eta^* = \eta^{**}$, i.e. η^* is a fixed point.

b) Suppose $\eta^* = \eta^{**}$, then (4.11) and (4.12) imply (4.10). ■

<u>REMARK 1</u>: According to Theorem 3.1 the "local" and "global" variants of the IDeC yield identical results, when applied to *linear* problems. This is of course not true for *nonlinear* problems, but Theorem 4.1 and Theorem 4.2 show that in the nonlinear case both variants have the same collocation solution as fixed point.

<u>REMARK 2</u>: From (4.2), it follows that we can construct for any "$s_{i,k}$-grid" of the IDeC-methods a corresponding "$t_{i,k}$-grid" of the collocation methods. Unfortunately the reverse is not true, e.g. for equidistant collocation nodes

$$\xi_k = k/(m+1), \quad k = 1(1)m,$$

there exists no corresponding "$s_{i,k}$-grid" for the box-scheme, which satisfies (4.2). Gauss-Legendre points with m even do not have a corresponding "$s_{i,k}$-grid" either, but for m odd an "$s_{i,k}$-grid" satisfying (4.2) may be found (see Fig. 1).

<u>Figure 1</u>

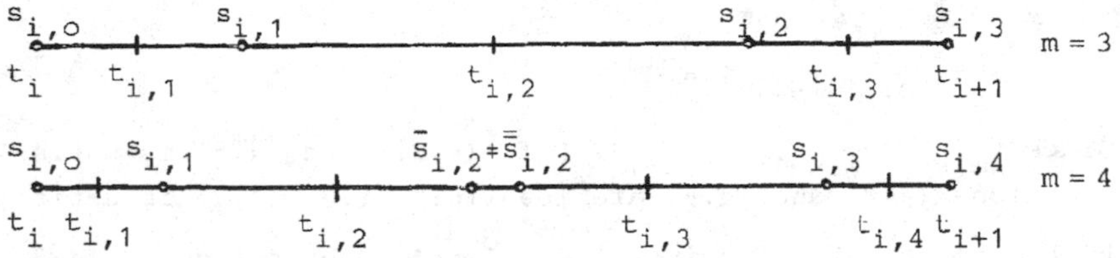

<u>4.1.2. IDeC-methods based on the scheme (3.4)</u>

The relation (4.2) becomes

$$t_{i,k} = s_{i,k-1}, \quad i = 0(1)I-1, \quad k = 1(1)m. \tag{4.13}$$

The corresponding collocation nodes satisfy

$$0 = \xi_1 < \ldots < \xi_m < 1. \tag{4.14}$$

Theorems corresponding to Theorem 4.1 and Theorem 4.2 will now hold for the IDeC-methods based on the scheme (3.4).

The "grid-restrictions" formulated in Remark 2 for the IDeC-methods based on the box-scheme do not apply in the present situation. There is a one-to one correspondence between "$t_{i,k}$-grids" and the "$s_{i,k}$-grids" (cf. (4.13)).

4.1.3. IDeC-methods based on the scheme (3.5)

Remarks analogous to those made for the scheme (3.4) apply to the scheme (3.5) with

$$t_{i,k} = s_{i,k}, \quad i = 0(1)I-1, \quad k = 1(1)m$$

and

$$0 < \xi_1 < \ldots < \xi_m = 1.$$

4.2. Initial value problems

All the results of Section 4.1 hold for IVPs, if the boundary condition

$$g(y(a),y(b)) = y(a) - y_a = 0 \tag{4.16}$$

is used. Collocation schemes for IVPs are only competitive for stiff systems of ODEs. For such problems methods with good stability properties are needed. Collocation schemes based on Gauss-Legendre points are known to be A-stable (cf. Wright [12]), and therefore, the IDeC based on the implicit midpoint rule (3.3)(with m odd) seems to be an appropriate scheme for solving stiff ODEs. Collocation schemes based on Radau points (with $\xi_m = 1$) are strongly A-stable (cf. Wright [12]) and therefore, the IDeC-methods based on the implicit Euler method (3.5) is perhaps an even more interesting scheme for solving stiff problems.

Up to now we have not examined whether the iterates *converge* to the fixed point. Consider the IDeC-methods based on the implicit Euler-method applied to stiff systems. This possibility has been investigated by Frank, Ueberhuber [6]. It is shown that, for equidistant nodes ($\xi_k = k/m$), very promising convergence results hold. Some of these results do not apply when the nodes are Radau points. As a consequence, it would appear that an IDeC-method based on the impl. Euler method on an *equidistant grid* is the preferred implementation.

If the IDeC is examined as a method for solving collocation equations, then J (the maximum number of IDeC-steps) is not determined by the asymptotic result (3.23). In this situation, the standard stopping criterion

$$\|\eta^j - \eta^{j-1}\| < \varepsilon$$

may be used.

4.3. Fixed points of IDeC-methods for other discretizations

In the previous sub-sections (4.1, 4.2), a relation between collocation schemes and the fixed point of the IDeC-methods based on methods (3.3), (3.4) and (3.5) was established. As a consequence, it is natural to examine whether the IDeC-methods based on arbitrary RK-methods always have "collocation fixed-points". In general, this question has to be answered in the negative, as a simple counter-example shows:

If the trapezoidal rule is used as the basic discretization method for the "global variant" of the IDeC, then the fixed point is characterized by

$$d_i^*(s_{i,k}) + d_i^*(s_{i,k+1}) = 0.$$

This fixed point is therefore only equivalent to a rather general weighted residual method, where instead of requiring the defects to vanish at any single grid point, a linear combination of the defects must vanish.

5. CONCLUSION

In this paper, iterative methods for solving collocation equations were introduced. Any step of the iterative process produces an approximation η^j which is usually more accurate than η^{j-1}. Compared with Newton's method for solving the collocation equations, the above strategy yields more information by which its implementation can be controlled (for example, step size control).

In Frank, Ueberhuber [6] the fact is discussed, that the effort necessary to perform the IDeC-steps is low compared with the effort necessary to solve the basic finite difference scheme ((3.3), (3.4) or (3.5)). Moreover, the structure of the equations (3.3), (3.4) and (3.5) is much simpler than the structure of the collocation equations. The IDeC-methods are therefore a more economical way for solving the collocation equations, than Newton's method. E.g. the application of the IDeC to stiff IVPs requires the solution of systems of non-linear equations of the same dimension n as the given problem, whereas the dimension of the collocation equations is $n \times m$, if a scheme with m collocation nodes is used.

A further advantage of the IDeC-methods for certain IVPs (with high stiffness) is the fast convergence of the approximations to the fixed point which corresponds to the solution of a collocation method. In some situations, an approximation to the fixed point which agrees with it to machine accuracy is obtained after *one* IDeC-step (see Frank, Ueberhuber [6]).

REFERENCES

[1] O. Axelsson
A Class of A-stable Methods
BIT Vol. 9 (1969), pp. 185 - 199

[2] F.H. Chipmann
Numerical Solution of Initial Value Problems using A-stable Runge-Kutta Processes
Dept. of A.A.C.S, Univ. of Waterloo,
Research Report, CSSR 2042, 1971

[3] C. de Boor, B. Swartz
Collocation at Gaussian Points
SIAM J. Num. Anal. Vol. 10(1973), pp. 582 - 606

[4] B.L. Ehle
On Padé Approximations to the Exponential Function and A-stable Methods for the Numerical Solution of Initial Value Problems
Dept. of A.A.C.S., Univ. of Waterloo
Research Report, CSSR 2010, 1969

[5] R. Frank, C.W.Ueberhuber
Iterated Defect Correction for Runge-Kutta Methods
Report Nr. 14/76, Inst. f. Num. Math.
Technical University of Vienna

[6] R. Frank, C.W. Ueberhuber
Iterated Defect Correction for the Efficient Solution of Stiff Systems of Ordinary Differential Equations
Report Nr. 17/76, Inst. f. Num. Math.
Technical University of Vienna

[7] R. Frank, J. Hertling, C.W. Ueberhuber
Iterated Defect Correction based on Estimates of the Local Discretization Error
Report Nr. 18/76, Inst. f. Num. Math.
Technical University of Vienna

[8] V.L. Pereyra
On Improving an Approximate Solution of a Functional Equation by Deferred Corrections
Num. Math. Vol. 8 (1966), pp. 376 - 391

[9] R.D. Russel, L.F. Shampine
A Collocation Method for Boundary Value Problems
Num. Math. Vol. 19 (1972), pp. 1 - 28

[10] H.J. Stetter
Economical Global Error Estimation, in
R.A. Willoughby (Ed.)
Stiff Differential Systems
Plenum Press, New York - London, 1974, pp. 245 - 258

[11] R. Weiss
The Application of Implicit Runge-Kutta and Collocation Methods to Boundary Value Problems
Math. Comp. Vol. 28 (1974), pp. 449 - 464

[12] K. Wright
Some Relationships between Implicit Runge-Kutta, Collocation and Lanczos τ Methods, and their Stability Properties
BIT Vol. 10 (1970), pp. 217 - 227

[13] P.E. Zadunaisky
A Method for the Estimation of Errors Propagated in the Numerical Solution of a System of Ordinary Differential Equations, in
Proc. Intern. Astron. Union
Symposium No. 25, Thessaloniki 1964
Academic Press, New York, 1966

VERALLGEMEINERTE RUNGE-KUTTA VERFAHREN ZUR LOESUNG STEIFER DIFFERENTIALGLEICHUNGSSYSTEME

A. Friedli
Seminar fuer Angewandte Mathematik
Eidgenössische Technische Hochschule
ETH-Zentrum, CH-8092 Zuerich

Ausgehend von gegebenen, expliziten Runge-Kutta Verfahren werden durch einen Ansatz mit unbekannten Koeffizienten verallgemeinerte Runge-Kutta Verfahren hergeleitet, die sich für steife Differentialgleichungssysteme eignen. Dazu wird die Jacobimatrix J des Systems, oder eine Näherung dazu, sowie deren Exponentialmatrix benötigt. Ist J für jeden Integrationsschritt verfügbar, kann die Ordnung der Verfahren um 1 erhöht werden. Die Verfahren lösen Systeme $y'(x)=Jy(x)+h(x)$, wo h ein Polynom gewissen Grades ist, exakt. Es werden Beispiele von verallgemeinerten Runge-Kutta Verfahren gegeben.

1 EINLEITUNG

Gegeben sei das Anfangswertproblem

(1.1) $y'(x) = f(x,y), \quad y(x_0) = y_0$

für ein System von gewöhnlichen Differentialgleichungen. Während sich für nichtsteife Systeme das Schwergewicht in der letzten Zeit auf die Umsetzung von bekannten numerischen Methoden zur Lösung von (1.1) in zuverlässige Computerroutinen konzentriert hat, scheint für steife Systeme die Suche nach geeigneten Methoden noch nicht abgeschlossen zu sein.

Die in dieser Arbeit hergeleiteten verallgemeinerten Runge-Kutta Verfahren (VRK Verfahren) fanden ihren Anstoss in einer von Rutishauser (1967) entwickelten ALGOL Prozedur, die im 2. Abschnitt beschrieben wird (siehe auch Rutishauser(1976)). Seine Methode verlangt im Prinzip die Verfügbarkeit der Jacobimatrix des Systems (1.1), respektive eine Näherung T an diese, und auch die Bildung deren Exponentialmatrix. Die Schwierigkeit der Berechnung der letzteren resultierte für die Prozedur in der Einschränkung von T auf Diagonalmatrizen.

Als weitere Beiträge auf dem Gebiet der Verallgemeinerung von Runge-Kutta Verfahren seien die Arbeiten von Lawson(1967), Ehle und Lawson (1975), Van der Houwen (1973) sowie Verwer (1975) erwähnt, die für die Exponentialmatrix geeignete Approximationen verwenden.

Die vorliegende Arbeit berichtet im 3. Abschnitt über die Verallgemeinerung von bekannten Runge-Kutta (RK) Verfahren zu VRK Verfahren. Aus der vorausgesetzten Verfügbarkeit der Jacobimatrix des Systems (1.1) an gewissen Stellen während der Integration kann im Hinblick auf eine Erhöhung der Ordnung der Verfahren Vorteil gezogen werden. Zusätzlich sollen die Verfahren Systeme der Form $y'(x) = Ay(x) + h(x)$, wo $h(x)$ ein Polynom ist, exakt lösen, wenn vorausgesetzt wird, dass die Exponentialmatrix genau berechnet wird. Die hergeleiteten Verfahren können aber auch mit (z.B. rationalen) Approximationen von angemessener Genauigkeitsordnung an die Exponentialmatrix Verwendung finden. Einige Beispiele von VRK Verfahren werden im 4. Abschnitt gegeben. Im 5. Abschnitt ist ein Beispiel für die Verallgemeinerung von eingebetteten RK Verfahren aufgeführt. Eingebettete Verfahren liefern ohne grossen Mehraufwand eine Schätzung des lokalen Diskretisationsfehlers, die für die Schrittweitensteuerung verwendet werden kann. Im 6. Abschnitt schliessen sich noch einige Bemerkungen über die Berechnung der Exponentialmatrix an.

Auf die Art und Weise der Umsetzung der beschriebenen Verfahren in Computerprogramme sowie auf numerische Erfahrungen soll in einer späteren Arbeit eingegangen werden. Ausführliche numerische Tests mit einem auf dem klassischen Runge-Kutta Verfahren 4. Ordnung basierenden VRK Verfahren wurden von Ehle (1974) sowie Enright et al (1975) durchgeführt.

2 BEISPIEL EINES VRK VERFAHRENS

Für die numerische Integration von (1.1) gehen wir aus von einem RK Verfahren 2. Ordnung (Verfahren von Heun):

$$y_1^* = y_n, \qquad k_1 = f(x_n, y_n)$$

$$y_2^* = y_n + hk_1, \qquad k_2 = f(x_n + h, y_2^*)$$

$$y_{n+1} = y_n + \frac{h}{2}(k_1+k_2).$$

Das für die erste Stufe verwendete Verfahren $y_2^* = y_n+hk_1$ (Euler) ist exakt für das auf dem Intervall $[x_n, x_n + h]$ definierte Problem

$$(2.1) \qquad \bar{y}'(x) = k_1, \qquad \bar{y}(x_n) = y_n.$$

Ersetzen wir $f(x,y)$ in (1.1) durch $Ty(x)+g(x,y)$, wobei T eine auf $[x_n, x_n+h]$ konstante Matrix bedeutet, lautet unser Differentialgleichungssystem

$$(2.2) \qquad y'(x) - Ty(x) = g(x,y), \qquad y(x_o) = y_o,$$

und das dem Problem (2.1) entsprechende Problem wird zu

$$\bar{y}'(x) - T\bar{y}(x) = g_1, \quad \bar{y}(x_n) = y_n; \quad (g_1 := g(x_n, y_n)).$$

Analytische Integration liefert

$$\bar{y}(x) = e^{(x-x_n)T} y_n + T^{-1}(e^{(x-x_n)T}-I)g_1,$$

was für $x = x_n+h$ auf

$$\bar{y}_2^* = e^{hT} y_n + h(hT)^{-1}(e^{hT}-I)g_1,$$

und mit den Definitionen

$$E_o(hT) := e^{hT}, \qquad E_1(hT) := (hT)^{-1}(e^{hT}-I)$$

auf

$$\bar{y}_2^* = E_o(hT)y_n + hE_1(hT)g_1 \quad \text{führt.}$$

Analog löst $y_{n+1} = y_n + \frac{h}{2}(k_1+k_2)$ das Problem

$$\bar{y}'(x) = k_1 + \frac{x-x_n}{h}(k_2-k_1), \quad \bar{y}(x_n) = y_n$$

exakt, das für das Differentialgleichungssystem (2.2) die folgende Form annimmt:

$$\bar{y}'(x) - T\bar{y}(x) = g_1 + \frac{x-x_n}{h}(g_2-g_1), \qquad \bar{y}(x_n) = y_n; \quad (g_2 := g(x_n+h, \bar{y}_2^*)).$$

Analytische Integration und die Substitution $x = x_n + h$ sowie die Definition $E_2(hT) := 2(hT)^{-2}(e^{hT}-I-hT)$ ergibt $\bar{y}_{n+1}$ (d.h. y_{n+1} in (2.3)).

Zusammenfassend haben wir also das folgende Verfahren hergeleitet:

$$(2.3)\quad \begin{cases} y_1^* = y_n, & g_1 = f(x_n, y_1^*) - Ty_1^* \\ y_2^* = E_0(hT)y_n + hE_1(hT)g_1, & g_2 = f(x_n+h, y_2^*) - Ty_2^* \\ y_{n+1} = E_0(hT)y_n + \frac{h}{2}((2E_1(hT) - E_2(hT))\, g_1 + E_2(hT)\, g_2). \end{cases}$$

Man sieht leicht, dass für $T = 0$ das Verfahren (2.3) in das Heusche Verfahren übergeht, vom dem wir ja ausgegangen sind. Das Verfahren wird sich um so besser für steife Differentialgleichungssysteme eignen, je näher die Matrix T der (momentanen) Jacobimatrix des Systems (1.1) gewählt wird. Das Problem bei der Auswertung von (2.3) besteht in der Berechnung der Exponentialmatrix $E_0(hT) = e^{hT}$ und deren verwandten Matrizen $E_1(hT)$ und $E_2(hT)$. In der von Rutishauser (1967) geschriebenen ALGOL Prozedur wurde denn auch T als Diagonalmatrix vorausgesetzt.

3 HERLEITUNG VON VRK VERFAHREN

a) RK Verfahren

Für die Herleitung von VRK Verfahren gehen wir immer aus von einem expliziten RK Verfahren mit s Stufen, das wie folgt charakterisiert ist:

$$(3.1)\quad \begin{cases} \left.\begin{array}{l} y_i^* = y_n + h\sum_{j=1}^{i-1} a_{ij}k_j \\ k_i = f(x_n+c_ih, y_i^*) \end{array}\right\} \quad (i=1,\ldots s) \\ y_{n+1} = y_n + h\sum_{i=1}^{s} b_ik_i \end{cases}$$

Das Verfahren ist gegeben durch die Koeffizienten c_i, a_{ij}, b_i, die mit folgendem Schema dargestellt werden können:

$$\begin{array}{c|ccccc} c_1 = 0 & & & & & \\ c_2 & a_{21} & & & & \\ c_3 & a_{31} & a_{32} & & & \\ . & . & . & . & & \\ c_s & a_{s1} & a_{s2} & . & a_{ss-1} & \\ \hline & b_1 & b_2 & . & b_{s-1} & b_s \end{array}$$

Diese Koeffizienten bestimmen sich aus den Bedingungsgleichungen, die sich aus dem Vergleich der Taylorentwicklung um x_n der exakten Lösung von $y'(x)=f(x,y)$, $y(x_n) = y_n$ mit der Entwicklung von y_{n+1} nach Potenzen von h ergeben. Für den mit dieser Technik verbundenen Mechanismus sei auf Butcher (1963) oder Henrici (1962/68) verwiesen.

Die (lokale) Ordnung q des Verfahrens ergibt sich aus der grössten Potenz von h, bis zu der die beiden Entwicklungen übereinstimmen. Wir fassen die Ordnungen q_i der y_i^* (i=2, ... s) und die Ordnung q von y_{n+1} zusammen in $Q:=(q_2, \dots q_s, q)$. Ein RK Verfahren der Ordnung q bezeichnen wir mit RK-q.

Das RK Verfahren besitze die Polynomordnung p, falls es die exakte Lösung für Systeme $y'(x)=h(x)$ liefert, wo h(x) ein beliebiges Polynom vom Grad kleiner oder gleich p ist. Die Polynomordnungen p_i der y_i^* (i=2, ... s) und p von y_{n+1} seien mit $P:=(p_2, \dots p_s, p)$ dargestellt.

b) VRK Verfahren

Basierend auf einem gegebenen RK Verfahren (3.1) machen wir den folgenden Ansatz für ein VRK Verfahren:

$$(3.2)\quad \begin{cases} \left.\begin{aligned} y_i^* &= E_o(c_i hT)y_n + h\sum_{j=1}^{i-1} A_{ij}(c_i hT)g_j \\ g_i &= f(x_n+c_i h, y_i^*) - Ty_i^* \end{aligned}\right\} & (i=1, \dots s) \\ y_{n+1} = E_o(hT)y_n + h\sum_{i=1}^{s} B_i(hT)g_i \end{cases}$$

$c_1 = 0$					
c_2	$A_{21}(c_2hT)$				
c_3	$A_{31}(c_3hT)$	$A_{32}(c_3hT)$			
.	.	.	.		
c_s	$A_{s1}(c_shT)$	$A_{s2}(c_shT)$	.	$A_{ss-1}(c_shT)$	
	$B_1(hT)$	$B_2(hT)$	.	$B_{s-1}(hT)$	$B_s(hT)$

Die Matrizen A_{ij} und B_i sind angesetzt als

$$A_{ij}(c_i hT) = \sum_{m=1}^{ma} \lambda_{ijm} E_m(c_i hT) \qquad (i=2, \dots s; j=1, \dots i-1)$$

$$B_i(hT) = \sum_{m=1}^{mb} \mu_{im} E_m(hT) \qquad (i=1, \dots s),$$

und die Matrizen E_m sind durch

$$(3.3)\quad E_o(Z) := e^Z, \qquad E_m(Z) := mZ^{-1}(E_{m-1}(Z)-I) \qquad (m=1, 2, \dots)$$

definiert, wobei auch die folgenden Darstellungen gelten:

$$(3.4)\quad E_m(Z) = \sum_{k=0}^{\infty} \frac{1}{(m+1)_k} Z^k \qquad (m = 0, 1, 2, \dots).$$

$$(\ (n)_k := n(n+1) \dots (n+k-1), \qquad (n)_o := 1)$$

$$(3.5)\qquad E_m(hZ) = \frac{m}{h^m} e^{hZ} \int_0^h e^{-tZ} t^{m-1} dt \qquad (m=1,\ 2,\ \ldots).$$

Beziehung (3.4) eingesetzt in die Ausdrücke für A_{ij} und B_i ergibt

$$A_{ij}(c_i hT) = \sum_{k=0}^{\infty} \alpha_{ijk}\, c_i^k\, h^k T^k$$

$$B_i(hT) = \sum_{k=0}^{\infty} \beta_{ik} h^k T^k, \quad \text{wobei}$$

$$(3.6)\qquad \alpha_{ijk} := \sum_{m=1}^{ma} \frac{\lambda_{ijm}}{(m+1)_k}, \qquad \beta_{ik} := \sum_{m=1}^{mb} \frac{\mu_{im}}{(m+1)_k}$$

c) Bedingungsgleichungen für VRK-q Verfahren

Wie bei der Herleitung von RK Verfahren sollen nun die Koeffizienten λ_{ijm} und μ_{im} so bestimmt werden, dass die Entwicklung von y_{n+1} nach Potenzen von h bis zur Ordnung q mit der Taylorentwicklung der exakten lokalen Lösung um x_n übereinstimmt. Dabei ist die Anzahl der Koeffizienten λ und μ nicht zum voraus festgelegt, d.h. die Summationsgrenzen ma und mb sind vorerst noch nicht bestimmt. Sie werden so gewählt werden müssen, dass die Bedingungsgleichungen erfüllt werden können.

Das VRK Verfahren (3.2) soll für beliebige Matrizen T gelten. Wie man sofort sieht, reduziert sich (3.2) für $T = 0$ zu einem RK Verfahren, das mit unserem vorgegebenen Verfahren (3.1) identisch sein soll. Wegen $A_{ij}(0) = \alpha_{ijo} I$ und $B_i(0) = \beta_{io} I$ ergeben sich die ersten Bedingungsgleichungen als

$$\alpha_{ijo} = a_{ij} \qquad (i=2, \ldots s;\ j = 1, \ldots i-1)$$
$$\beta_{io} = b_i \qquad (i=1, \ldots s)$$

(Obwohl die eigentlichen unbekannten Koeffizienten die λ und die μ sind, werden wir die Bedingungsgleichungen in den α und β aufstellen; die Beziehungen sind gegeben durch (3.6).)

d) Bedingungsgleichungen für VRK-q^+ Verfahren

Es ist klar, dass für beliebige T die Ordnung von (3.2) nicht grösser sein kann als für (3.1). Hingegen können wir versuchen, die Ordnung um 1 zu erhöhen, wenn wir für T die Jacobimatrix J_n ($= \partial f(x,y)/\partial y$ an der Stelle (x_n, y_n)) verwenden. Wir bezeichnen die Ordnung eines VRK Verfahrens mit q^+, wenn es die Ordnung q besitzt für beliebige T, und die Ordnung q+1 für $T=J_n$; wir benützen die Notation VRK-q^+.

In Tabelle 1 sind die Bedingungsgleichungen für VRK Verfahren für die Ordnungen 1 (resp. 1^+) bis 4 (resp. 4^+) aufgeführt. Die erste Summation erstreckt sich immer über i von 1 bis s, die zweite über j von 1 bis i-1, die dritte über k von 1 bis j-1 usw. Um die Ordnung q (resp. q^+) zu erreichen, müssen alle Gleichungen bis zum

TABELLE 1

Ordnung	RK - Gleichungen	VRK - Gleichungen
1	$\Sigma b_i = 1$	$\beta_{i0} = b_i \quad (i=1,\ldots s)$
1^+		$\Sigma\beta_{i1} = 1/2$
2	$\sum_{j=1}^{i-1} a_{ij} = c_i \quad (i-2,\ldots s)$	$\alpha_{ij0} = a_{ij} \quad (i=2,\ldots s;\ j=1,\ldots i-1)$
	$\Sigma b_i c_i = 1/2$	
2^+	$\Sigma b_i c_i^2 = 1/3$	$\Sigma\beta_{i2} = 1/6$
3	$\Sigma b_i \Sigma a_{ij} c_j = 1/6$	$\Sigma\beta_{i1} c_i = 1/6$
		(*) $\sum_{j=1}^{i-1} \alpha_{ij1} = c_i/2 \quad (i=2,\ldots s)$
3^+	$\Sigma b_i c_i^3 = 1/4$	$\Sigma\beta_{i1} c_i^2 = 1/12$
		$\Sigma\beta_{i3} = 1/24$
		(x)
4	$\Sigma b_i c_i \Sigma a_{ij} c_j = 1/8$	$\Sigma\beta_{i1} \Sigma a_{ij} c_j = 1/24$
	$\Sigma b_i \Sigma a_{ij} c_j^2 = 1/12$	$\Sigma\beta_{i2} c_i = 1/24$
	$\Sigma b_i \Sigma a_{ij} \Sigma a_{jk} c_k = 1/24$	(**) $\Sigma b_i c_i^2 \Sigma\alpha_{ij2} = 1/24$
		$\Sigma b_i c_i \Sigma\alpha_{ij1} c_j = 1/24$
4^+	$\Sigma b_i c_i^4 = 1/5$	$\Sigma\beta_{i1} c_i^3 = 1/20$
	$\Sigma b_i c_i \Sigma a_{ij} c_j^2 = 1/15$	$\Sigma\beta_{i2} c_i^2 = 1/60$
		$\Sigma\beta_{i4} = 1/120$
		(xx) $\Sigma b_i c_i^3 \Sigma\alpha_{ij2} = 1/30$

TABELLE 2

Ord-nung	RK - Gleichungen	VRK - Gleichungen
4^+		(**) $\Sigma \alpha_{ij2} = c_i/6$ (i=2,...s) (xx) weglassen
	$\Sigma b_i c_i^2 \Sigma a_{ij} c_j = 1/10$	$\Sigma \beta_{i1} \Sigma a_{ij} c_j^2 = 1/60$
	$\Sigma b_i c_i \Sigma a_{ij} \Sigma a_{jk} c_k = 1/30$	$\Sigma \beta_{i1} c_i \Sigma a_{ij} c_j = 1/40$
	$\Sigma b_i (\Sigma a_{ij} c_j)^2 = 1/20$	$\Sigma \beta_{i1} \Sigma a_{ij} \Sigma a_{jk} c_k = 1/120$
	$\Sigma b_i \Sigma a_{ij} c_j^3 = 1/20$	$\Sigma \beta_{i2} \Sigma a_{ij} c_j = 1/120$
	$\Sigma b_i \Sigma a_{ij} c_j \Sigma a_{jk} c_k = 1/40$	$\Sigma \beta_{i3} c_i = 1/120$
	$\Sigma b_i \Sigma a_{ij} \Sigma a_{jk} c_k^2 = 1/60$	($) $\Sigma \beta_{i1} c_i \Sigma \alpha_{ij1} c_j = 1/120$
	$\Sigma b_i \Sigma a_{ij} \Sigma a_{jk} \Sigma a_{kl} = 1/120$	$\Sigma b_i \Sigma a_{ij} c_j \Sigma \alpha_{jkl} c_k = 1/120$
		$\Sigma b_i c_i \Sigma \alpha_{ij1} c_j^2 = 1/60$
		$\Sigma b_i c_i \Sigma \alpha_{ij1} \Sigma a_{jk} c_k = 1/120$
		$\Sigma b_i c_i^2 \Sigma \alpha_{ij2} c_j = 1/120$
		$\Sigma b_i c_i^2 \Sigma \alpha_{ij1} c_j = 1/30$
5		$\Sigma b_i c_i^3 \Sigma \alpha_{ij3} = 1/120$
	$\Sigma b_i c_i^5 = 1/6$	$\Sigma \beta_{i1} c_i \Sigma a_{ij} c_j^2 = 1/90$
	$\Sigma b_i c_i \Sigma a_{ij} c_j^3 = 1/24$	$\Sigma \beta_{i1} c_i^4 = 1/30$
	$\Sigma b_i c_i^2 \Sigma a_{ij} c_j^2 = 1/18$	$\Sigma \beta_{i2} c_i^3 = 1/120$
		$\Sigma \beta_{i3} c_i^2 = 1/360$
		$\Sigma \beta_{i5} = 1/720$
		$\Sigma b_i c_i^2 \Sigma \alpha_{ij1} c_j^2 = 1/72$
5^+		$\Sigma b_i c_i^4 \Sigma \alpha_{ij3} = 1/144$

entsprechenden Strich erfüllt werden. Man beachte, dass ein VRK Verfahren der Ordnung q^+ gegenüber einem solchen der Ordnung q nicht nur zusätzliche VRK Gleichungen, sondern auch zusätzliche RK Gleichungen befriedigen muss. Es wird sich also nicht jedes gegebene RK-q Verfahren auf ein VRK-q^+ Verfahren verallgemeinern lassen. Falls ein RK-q Verfahren die zusätzlichen RK Gleichungen erfüllt, deuten wir das (in Analogie zu oben) an mit der Bezeichnung RK-q^+.

Die Gleichungen (*) sind Annahmen, die bewirken, dass keine quadratischen Gleichungen in den α und/oder β auftreten. Für Ordnungen bis und mit 3^+ können diese Annahmen jedoch weggelassen werden; in diesem Fall muss (*) ersetzt werden durch

$$\Sigma\, b_i c_i \;\; \Sigma\, \alpha_{ijl} = 1/6$$

und an die Stelle (x) muss

$$\Sigma\, b_i c_i^2 \;\Sigma\, \alpha_{ijl} = 1/8$$

gesetzt werden.

Tabelle 2 setzt die Gleichungen fort bis zur Ordnung 5^+. Dabei muss Gleichung (**) in Tabelle 1 ersetzt werden durch die Annahmen

$$\sum_{j=1}^{i-1} \alpha_{ij2} = c_i/6 \qquad (i = 2, \ldots s)$$

und Gleichung (xx) weggelassen werden. Trotz den beiden Annahmen bleibt die quadratische Gleichung ($) übrig. Wir haben die α-Gleichungen ohne ($) gelöst und die (nun lineare) Gleichung ($) bei den β-Gleichungen berücksichtigt.

e) <u>Bedingungsgleichungen für Polynomordnung</u> $P_{VRK} = (p_2, \ldots p_s, p)$

Schliesslich erhalten wir weitere Bedingungsgleichungen durch die Forderung, dass das VRK Verfahren (3.2) Differentialgleichungssysteme der Form

$$(3.7) \qquad y'(x) = Ay(x) + h(x), \qquad y(x_o) = y_o,$$

wo h(x) ein beliebiges Polynom vom Grad kleiner oder gleich p ist, für T = A exakt löst ; wir sagen, das VRK Verfahren besitzt die <u>Polynomordnung</u> p. Die folgenden Ueberlegungen genügen für die Herleitung dieser Bedingungsgleichungen.

Die exakte Lösung von

$$(3.8) \qquad y'(x) = Ty(x) + x^{k-1} r, \; y(0) = y_o \, (k > 0) \quad (r := (1, \ldots 1)^T)$$

lautet

$$y(x) = e^{xT} y_o + e^{xT} \int_0^x e^{-tT} t^{k-1} dt r,$$

was für x = h und unter Verwendung von (3.5)

$$y(x) = E_o(hT) y_o + \frac{h^k}{k} E_k(hT) r \qquad \text{ergibt.}$$

Andererseits liefert das VRK Verfahren (3.2) für das Problem (3.8)

$$y_1 = E_0(hT)y_0 + h \sum_{i=1}^{s} B_i(hT)\, c_i^{k-1} h^{k-1} r.$$

Der Vergleich liefert die Bedingungen

$$\sum_{i=1}^{s} B_i(hT)c_i^{k-1} = \frac{1}{k} E_k(hT) \qquad (k=1, \ldots p+1).$$

Setzen wir $B_i(hT) = \sum_{m=1}^{mb} \mu_{im} E_m(hT)$ ein, ergeben sich die folgenden Bedingungsgleichungen:

$$(3.9) \qquad \sum_{m=1}^{mb} \sum_{i=1}^{s} \delta_{mu} c_i^{k-1} \mu_{im} = \frac{1}{u} \delta_{ku} \quad (k=1, \ldots p+1; \quad u = 1, \ldots mb).$$

$(\delta_{ku} := 1$ für $k=u$, $\delta_{ku} := 0$ sonst.)

In analoger Weise führt die Forderung, dass ein y_i^* in (3.2) das System (3.7) für beliebige Polynome $h(x)$ mit Grad kleiner oder gleich p_i $(i = 2, \ldots s)$ exakt lösen soll, auf die folgenden Bedingungen:

$$(3.10) \qquad \sum_{m=1}^{ma} \sum_{i=2}^{s} \sum_{j=1}^{i-1} \delta_{iu} \delta_{mv} c_j^{k-1} \lambda_{ijm} = \frac{1}{v} \delta_{kv} c_u^v$$

$$(u=2, \ldots s; \quad k = 1, \ldots p_u+1; \quad v = 1, \ldots ma).$$

Die Bedingungen (3.9) und (3.10) sind direkt in den Unbekannten μ und λ ausgedrückt, da sie sich aus exakt zu erfüllenden Forderungen ergeben und nicht nur bis zu einer gewissen Ordnung gelten sollen.

f) Auflösung der Gleichungssysteme

Wie oben erwähnt, müssen für die λ und μ zwei separate lineare Gleichungssysteme aufgelöst werden. Die oberen Summationsgrenzen ma und mb hängen ab von der Wahl der Anzahl Unbekannten λ_{ijm} und μ_{im}. (Die Anordnung der Unbekannten geschieht - am Beispiel s = 3 - wie folgt:

$(\lambda_{211}, \lambda_{311}, \lambda_{321}; \lambda_{212}, \lambda_{312}, \lambda_{322}; \ldots)$ und $(\mu_{11}, \mu_{21}, \mu_{31}; \mu_{12}, \mu_{22}, \mu_{32}; \ldots)$.) Aus den Gleichungen (3.10) und (3.9) ergeben sich die Forderungen

$$(3.11) \quad ma \geq \mathrm{Max}_{2 \leq i \leq s}(p_i+1), \qquad mb \geq p+1.$$

Da viele der Bedingungsgleichungen linear abhängig sind, ist es aber nicht nötig, gleich viele Unbekannte U wie Gleichungen G einzuführen: Zuerst wird U auf die Anzahl G der sich aus Tabellen 1 und 2 ergebenden Gleichungen festgelegt. Eventuell muss U dann erhöht werden, damit Bedingung (3.11) erfüllt wird. Schliesslich wird G um die Anzahl der sich aus (3.10) resp. (3.9) ergebenden Gleichungen erweitert. Für Gleichungssysteme, die nicht den vollen Rang U aufwiesen, wurden einfach die als freie Parameter auftretenden Unbekannten Null gesetzt. In einigen Fällen wurde auch der Wert gewisser "Unbekannten" zum voraus festgelegt; wenn z.B. $b_2 = 0$, kann $\mu_{2m}=0$ (m=1, ... mb) gefordert werden. Die exakte Auflösung der Gleichungssysteme wurde mit der an der ETH Zürich zur Verfügung stehenden Formelsprache SYMBAL (Engeli (1975)) durchgeführt.

4 VRK-q^+ VERFAHREN (q = 1, 2, 3, 4)

a) Ausgehend vom RK-1^+ Verfahren

$$y_{n+1} = y_n + hf(x_n, y_n) \qquad \text{(Euler)}$$

ergibt sich sofort das VRK-1^+ Verfahren

$$y_{n+1} = E_0(hT)y_n + hE_1(hT)(f(x_n,y_n)-Ty_n).$$

Beide Verfahren besitzen die Polynomordnung 0.

b) Das einzige RK-2^+ Verfahren (mit 2 Stufen) lautet:

$$\begin{array}{c|cc} 0 & & \\ \frac{2}{3} & \frac{2}{3} & \\ \hline & \frac{1}{4} & \frac{3}{4} \end{array} \qquad Q_{RK} = (1^+, 2^+) \quad P_{RK} = (0, 2)$$

Es lässt sich zum folgenden VRK-2^+ Verfahren erweitern:

$$\begin{array}{c|cc} 0 & & \\ \frac{2}{3} & \frac{2}{3}E_1(\frac{2}{3}hT) & \\ \hline & E_1(hT) - \frac{3}{4}E_2(hT) & \frac{3}{4}E_2(hT) \end{array} \qquad Q_{VRK} = (1^+, 2^+) \quad P_{VRK} = (0, 1)$$

Ein VRK-2^+ Verfahren mit Polynomordnung 2 kann nicht gefunden werden, da sich Widersprüche in den Bedingungsgleichungen ergeben.

c) Alle für eine Verallgemeinerung zu einem VRK-3^+ Verfahren in Frage kommenden 3-stufigen RK-3^+ Verfahren lassen sich in der folgenden einparametrigen Schar darstellen:

$x \neq \frac{2}{3}, \frac{3}{4}$:

$c_1 = 0,\quad c_2 = x,\quad c_3 = (4c_2-3)/(3c_2-2)/2;$

$b_3 = (3c_2-2)/(c_2-c_3)/c_3/6,$

$b_2 = (3c_3-2)/(c_3-c_2)/c_2/6,\qquad b_1 = 1-(b_2+b_3);$

$a_{21}=c_2,\quad a_{32}=1/c_2/b_3/6,\qquad a_{31} = c_3-a_{32};$

$Q_{RK} = (1^+, 1^+, 3^+),\qquad P_{RK} = (0, 0, 3).$

Für $x = \frac{1}{2}$ ergibt sich beispielsweise:

0			
$\frac{1}{2}$	$\frac{1}{2}$		
1	-1	2	
	$\frac{1}{6}$	$\frac{2}{3}$	$\frac{1}{6}$

Die von Null verschiedenen Koeffizienten λ_{ijm} und μ_{im} der einparametrigen Schar von VRK-3^+ Verfahren lauten:

$$x \neq \frac{2}{3}, \frac{3}{4}:$$

$$\lambda_{211} = x$$

$$\lambda_{311} = -(3-16x+30x^2-18x^3)(3-4x)/(2-3x)^3/x/4$$

$$\lambda_{321} = (3-8x+6x^2)(3-4x)/(2-3x)^3/x/4$$

$$\mu_{11} = 1, \qquad \mu_{12} = -3/(1-2x^2)/(3-4x)/x/2$$

$$\mu_{13} = 2(2-3x)/(3-4x)/x/3$$

$$\mu_{22} = (3-4x)/(3-8x+6x^2)/x/2$$

$$\mu_{23} = -2(2-3x)/(3-8x+6x^2)/x/3$$

$$\mu_{32} = -2x(2-3x)^2/(3-8x+6x^2)/(3-4x)$$

$$\mu_{33} = 4(2-3x)^2/(3-8x+6x^2)/(3-4x)/3$$

$$Q_{VRK} = (1^+, 1^+, 3^+), \qquad P_{VRK} = (0,0,2).$$

(Die Forderung nach der Polynomordnung 3 kann nicht erfüllt werden.) Das VRK-3^+ Verfahren für $x = \frac{1}{2}$ lautet beispielsweise:

0			
$\frac{1}{2}$	$\frac{1}{2}\ E_1(\frac{1}{2}\ hT)$		
1	$-E_1(hT)$	$2E_1(hT)$	
	$(E_1 - \frac{3}{2}E_2 + \frac{2}{3}E_3)(hT)$	$(2E_2 - \frac{4}{3}E_3)(hT)$	$(-\frac{1}{2}E_2 + \frac{2}{3}E_3)(hT)$

d) Beim Versuch der Verallgemeinerung eines RK-4^+ Verfahrens auf ein VRK-4^+ Verfahren findet man, dass dies nur möglich ist mit der Stufenanzahl s=5, da die 13 RK-Gleichungen (in den 13 Unbekannten c_i (i=2, 3, 4), a_{ij} (i=2,3,4; j=1, ... i-1), b_i (i=1, 2, 3, 4)) für s = 4 keine Lösung besitzen (siehe Tabelle 1). (Natürlich können viele VRK-4 Verfahren mit s = 4 gefunden werden.) Als ein RK-4^+ Verfahren mit s = 5 stellt sich zum Beispiel das von Scraton (1965) gegebene heraus:

$$
\begin{array}{c|ccccc}
0 & & & & & \\
\frac{2}{9} & \frac{2}{9} & & & & \\
\frac{1}{3} & \frac{1}{12} & \frac{1}{4} & & & \\
\frac{3}{4} & \frac{69}{128} & -\frac{243}{128} & \frac{135}{64} & & \\
\frac{9}{10} & -\frac{621}{2000} & \frac{729}{400} & -\frac{1377}{1250} & \frac{306}{625} & \\
\hline
 & \frac{17}{162} & 0 & \frac{81}{170} & \frac{32}{135} & \frac{250}{1377}
\end{array}
$$

$Q_{RK} = (1^+, 2^+, 2^+, 2^+, 4^+)$, $\qquad P_{RK} = (0, 2, 2, 2, 4)$.

Ein auf diesem RK-4^+ Verfahren basierenden <u>VRK-4^+ Verfahren</u> ist das folgende:

$$
\begin{aligned}
&A_{21}(\tfrac{2}{9}\,hT) = \frac{2}{9} E_1, && A_{31}(\tfrac{1}{3}hT) = \frac{1}{3} E_1 - \frac{1}{4} E_2 \\
&A_{32}(\tfrac{1}{3}\,hT) = \frac{1}{4} E_2, && A_{41}(\tfrac{3}{4}\,hT) = \frac{231}{128} E_1 - \frac{81}{64} E_2 \\
&A_{42}(\tfrac{3}{4}\,hT) = -\frac{405}{128} E_1 + \frac{81}{64} E_2, && A_{43}(\tfrac{3}{4}\,hT) = \frac{135}{64} E_1 \\
&A_{51}(\tfrac{9}{10}hT) = \frac{189}{125} E_1 - \frac{729}{400} E_2, && A_{52}(\tfrac{9}{10}\,hT) = \frac{729}{400} E_2 \\
&A_{53}(\tfrac{9}{10}hT) = -\frac{1377}{1250} E_1, && A_{54}(\tfrac{9}{10}\,hT) = \frac{306}{625} E_1 \\
&B_1(hT) = E_1 - \frac{49}{18} E_2 + \frac{238}{81} E_3 - \frac{10}{9} E_4, && B_2(hT) = 0 \\
&B_3(hT) = \frac{729}{170} E_2 - \frac{594}{85} E_3 + \frac{54}{17} E_4 && \\
&B_4(hT) = -\frac{16}{5} E_2 + \frac{1184}{135} E_3 - \frac{16}{3} E_4 && \\
&B_5(hT) = \frac{250}{153} E_2 - \frac{6500}{1377} E_3 + \frac{500}{153} E_4 && \\
&Q_{VRK} = (1^+, 2^+, 2^+, 2^+, 4^+), && P_{VRK} = (0, 1, 1, 1, 3).
\end{aligned}
$$

Ein Verfahren mit $P_{VRK} = (0, 1, 1, 1, 2)$ erhält man, indem die obigen $B_m(hT)$ ersetzt werden durch:

$$
\begin{aligned}
&B_1(hT) = \frac{88}{81} E_1 - \frac{125}{54} E_2 + \frac{4}{3} E_3, \qquad B_2(hT) = 0 \\
&B_3(hT) = -\frac{21}{85} E_1 + \frac{531}{170} E_2 - \frac{12}{5} E_3 \\
&B_4(hT) = \frac{56}{135} E_1 - \frac{56}{45} E_2 + \frac{16}{15} E_3 \\
&B_5(hT) = -\frac{350}{1377} E_1 + \frac{200}{459} E_2 .
\end{aligned}
$$

Eine Lösung mit der Polynomordnung $p = 4$ wurde nicht gefunden. Auch treten Widersprüche auf bei der Forderung $P_{VRK} = (0, 2, 2, 2, p), p < 4$. Hingegen gibt es Lösungen für $P_{VRK} = (0, 1, 2, 2, p)$, $p < 4$.

5 EINGEBETTETE VRK VERFAHREN

Es sei (3.1) ein RK Verfahren der Ordnung $\bar{q}+1$. Stellt das Verfahren

$$\bar{y}_{n+1} = y_n + h \sum_{i=1}^{\bar{s}} \bar{b}_i k_i$$

- unter Verwendung derselben k_i (d.h. a_{ij}) wie in (3.1) - ein Verfahren der Ordnung $\bar{q}$ dar, spricht man von einem eingebetteten Verfahren. Für das eingebettete RK-$\bar{q}$ Verfahren ergibt sich so eine Abschätzung des lokalen Diskretisationsfehlers, die für eine automatische Schrittweitensteuerung Anwendung finden kann. Eingebettete RK Verfahren wurden zum Beispiel hergeleitet von Fehlberg (1969, 1970), Sarafyan (1966) und England (1969). Von den beiden letzteren Autoren wurde das folgende RK-4(5) Verfahren vorgeschlagen:

0						
$\frac{1}{2}$	$\frac{1}{2}$					
$\frac{1}{2}$	$\frac{1}{4}$	$\frac{1}{4}$				
1	0	-1	2			
RK-4	$\frac{1}{6}$	0	$\frac{2}{3}$	$\frac{1}{6}$		
$\frac{2}{3}$	$\frac{7}{27}$	$\frac{10}{27}$	0	$\frac{1}{27}$		
$\frac{1}{5}$	$\frac{28}{625}$	$-\frac{1}{5}$	$\frac{546}{625}$	$\frac{54}{625}$	$-\frac{378}{625}$	
RK-5	$\frac{1}{24}$	0	0	$\frac{5}{48}$	$\frac{27}{56}$	$\frac{125}{336}$

$Q_{RK} = (1^+,2,2,2,2,5)$, $\qquad P_{RK} = (0,1,1,1,1,4)$.

Für das eingebettete Verfahren gilt $\bar{q} = 4$ und $\bar{p} = 3$.

Eine mögliche Verallgemeinerung dieses Verfahrens zu einem VRK-4(5) Verfahren lautet wie folgt:

0						
$\frac{1}{2}$	$\frac{1}{2}E_1$					
$\frac{1}{2}$	$\frac{1}{2}E_1 - \frac{1}{4}E_2$	$\frac{1}{4}E_2$				
1	E_1-E_2	$-\frac{26}{25}E_1 + \frac{1}{25}E_2$	$\frac{26}{25}E_1 + \frac{24}{25}E_2$			
VRK-4	$E_1 - \frac{3}{2}E_2 + \frac{2}{3}E_3$	0	$2E_2 - \frac{4}{3}E_3$	$-\frac{1}{2}E_2 + \frac{2}{3}E_3$		
$\frac{2}{3}$	$\frac{127}{270}E_1 - \frac{19}{90}E_2$	$\frac{53}{135}E_1 - \frac{1}{45}E_2$	0	$-\frac{53}{270}E_1 + \frac{7}{30}E_2$		
$\frac{1}{5}$	$\frac{53}{625}E_1 - \frac{1}{25}E_2$	$\frac{-6}{25}E_1 + \frac{1}{25}E_2$	$\frac{546}{625}E_1$	$\frac{54}{625}E_1$	$-\frac{378}{625}E_1$	
VRK-5	B_1	B_2	B_3	B_4	B_5	B_6

$$B_1 = E_1 - \frac{15}{4} E_2 + \frac{14}{3} E_3 - \frac{15}{8} E_4, \qquad B_2 = 0, \qquad B_3 = 0$$

$$B_4 = \frac{1}{4} E_2 - \frac{13}{12} E_3 + \frac{15}{16} E_4, \qquad B_5 = -\frac{27}{28} E_2 + \frac{27}{7} E_3 - \frac{135}{56} E_4$$

$$B_6 = \frac{125}{28} E_2 - \frac{625}{84} E_3 + \frac{375}{112} E_4 .$$

$$Q_{VRK} = (1^+,2,2,2,2,5), \qquad \bar{q} = 4; \qquad P_{VRK} = (0,1,1,1,1,3), \ \bar{p} = 2.$$

Um dieses P_{VRK} zu erhalten, musste $A_{41} \neq 0$ akzeptiert werden ($a_{41}=0!$). Die Forderung nach $\bar{p} = 3$ kann nicht erfüllt werden und diejenige nach $p = 4$ nur, falls $B_2 \neq 0$ und $B_3 \neq 0$ zugelassen wird.

6 ZUR BERECHNUNG DER EXPONENTIALMATRIX

Die in einem VRK Verfahren (3.2) auftretenden Matrizen A_{ij} und B_i sind dargestellt als Linearkombinationen von gewissen Exponentialtermen E_m, die rekursiv durch (3.3) definiert sind. Für die Berechnung dieser E_m werden im allgemeinen eine genügend genaue rationale Approximation an E_o verwendet, und die Approximationen an die E_m ($m \geq 1$) gemäss (3.3) bestimmt. Eine Alternative wäre die direkte Approximation der A_{ij} und B_i. Schliesslich kann versucht werden, die E_m so genau wie möglich zu berechnen.

Eine der zahlreichen Möglichkeiten besteht darin, das vollständige Eigensystem der Matrix T zu bestimmen. Die zum Beispiel in EISPACK (Smith et al (1976)) vorhandenen FORTRAN Subroutinen erlauben für T die Zerlegung

$$T = Q\,R\,D(QR)^{-1},$$

wobei Q eine orthonormierte Matrix und R eine Rechtsdreiecksmatrix bedeuten. D ist "fast" eine Diagonalmatrix: Einem reellen Eigenwert ist ein Diagonalelement, einem konjugiert komplexen Eigenwertpaar eine 2x2 Matrix auf der Diagonalen zugeordnet. Für n= 3 also zum Beispiel (Eigenwerte c und $a \pm ib$)

$$D = \begin{pmatrix} a & b & 0 \\ -b & a & 0 \\ 0 & 0 & c \end{pmatrix}. \qquad \text{Es folgt}$$

$$E_m(hT) = QRE_m(hD)(QR)^{-1},$$

$$E_m(hD) = \begin{pmatrix} \mathrm{Re}(e_m(hz)) & \mathrm{Im}(e_m(hz)) & 0 \\ -\mathrm{Im}(e_m(hz)) & \mathrm{Re}(e_m(hz)) & 0 \\ 0 & 0 & e_m(hc) \end{pmatrix}$$

$$(6.1) \qquad z = a+ib, \qquad e_o(hz) = e^{hz}, \qquad e_m(hz) = \frac{m}{hz}\,(e_{m-1}(hz)-1), \qquad m = 1,2,\ldots$$

Rekursion (6.1) ist für $m \to \infty$ unstabil. Es zeigt sich aber, dass für kleine m und für $\mathrm{Re}(hz) \leq 0$ (was für die Eigenwerte von steifen Systemen der Fall ist), (6.1) stabil ist.

Es ist klar, dass für kleine $|hz|$ die $e_m(hz)$ in eine Reihe oder einen Kettenbruch um 0 entwickelt werden müssen.

Für die Anwendung auf die VRK Verfahren (3.2) soll auch berücksichtigt werden, dass z.B. die $B_i(hT)$ nie allein gebraucht werden, sondern dass diese immer mit einem Vektor v multipliziert werden:

$$B_i(hT)v = QR\tilde{D}(QR)^{-1}v; \qquad \tilde{D} := \sum_{m=1}^{mb} \mu_{im} E_m(hD).$$

$$B_i(hT)v = QR\tilde{D}x; \quad Rx = Q^T v.$$

Der Vektor x wird also durch ein einfaches Rückwärtseinsetzen berechnet. Die bei mehrfachen Eigenwerten möglichen Komplikationen bei der Bestimmung des Eigensystems von T werden sich in der grossen Konditionszahl der Rechtsdreiecksmatrix R ausdrücken. Da die hergeleiteten VRK-q Verfahren aber für beliebige T gelten, kann in einem solchen Fall das R so gestört werden, dass der Prozess des Rückwärtseinsetzens stabil verläuft. Diese Aenderung von R bedeutet eine Verwendung einer anderen Matrix T. Eine mathematisch präzisere Formulierung dieser heuristischen Ueberlegungen muss noch durchgeführt werden.

LITERATUR

Butcher, J.C. (1963). Coefficients for the Study of Runge-Kutta Integration Processes, J. Austr. Math. Soc. 3, 185-201.

Ehle, B. L. (1974). A Comparison of Numerical Methods for Solving Certain Stiff Ordinary Differential Equations, Dept. of Math. Report No. 70 (revised), Univ. of Victoria, Victoria B.C.

Ehle, B. L. and Lawson, J. D. (1975). Generalized Runge-Kutta Processes for Stiff Initial-Value Problems, J. Inst. Math. Appl. 16, No. 1, 11-21.

Engeli, M. (1975). SYMBAL Manual, FIDES Treuhandgesellschaft Zürich.

England, R. (1969). Error Estimates for Runge-Kutta Type Solutions to Systems of Ordinary Differential Equations, Comput. J. 12, 166-170.

Enright, W.H., Hull, T.E. and Lindberg, B. (1975). Comparing Numerical Methods for Stiff Systems of O.D.E.s, BIT 15, 10-48.

Fehlberg, E. (1969). Klassische Runge-Kutta-Formeln fünfter und siebenter Ordnung mit Schrittweiten-Kontrolle, Computing 4, 93-106.

Fehlberg, E. (1970). Klassische Runge-Kutta-Formeln vierter und niedriger Ordnung mit Schrittweiten-Kontrolle und ihre Anwendung auf Wärmeleitungsprobleme, Computing 6, 61-71.

Henrici, P. (1962/1968). Discrete Variable Methods in Ordinary Differential Equations, John Wiley, New York.

Lawson, J.D. (1967). Generalized Runge-Kutta Processes for Stable Systems with Large Lipschitz Constants, SIAM J. Numer. Anal., Vol. 4, No. 3, 372-380.

Rutishauser, H. (1967). ALGOL-Prozedur DAMINT. Institut für Angewandte Mathematik, ETH Zürich.

Rutishauser, H. (1976). (Herausgeber M. Gutknecht). Vorlesung über numerische Mathematik, §8.9, Birkhäuser Verlag, Basel.

Sarafyan, D. (1966) Error Estimates for Runge-Kutta Methods through Pseudo-Iterative Formulas. Dept. of Math. Technical Report No. 14,Louisiana State University in New Orleans.

Scraton, R.E. (1965). Estimation of the Truncation Error in Runge-Kutta and Allied Processes, Computer J. 7, 246-248.

Smith, B.T. et al (1976). Matrix Eigensystem Routines - EISPACK-Guide, 2. edition, Lecture Notes in Computer Science, Vol. 6, Springer, New York.

Van der Houwen, P.J. (1973). One-Step Methods with Adaptive Stability Functions for the Integration of Differential Equations,in Lecture Notes in Mathematics, Vol. 333, Springer, New York.

Verwer, J.G. (1975). S-Stability for Generalized Runge-Kutta Methods, Conference on Numerical Analysis, July 1975 in Dundee.

A FAST ITERATIVE METHOD FOR SOLVING POISSON'S EQUATION IN A GENERAL REGION

by W. Hackbusch

§ 1 Introduction

This paper discusses a fast method of solving difference equations, which approximate the solution u(x,y) of the boundary value problem

(1) $- \Delta u(x,y) = q(x,y) \qquad (x,y) \in G$

(2) $u(x,y) = r(x,y) \qquad (x,y) \in \partial G$

in an open and bounded region G. The system of the difference equations (e.g. obtained by the five-point-formula or by the Mehrstellenverfahren (cf. [3])) is solved iteratively. The computation of an approximation with accuracy ε requires $O(h^{-2}|\log \varepsilon|)$ operations (h: mesh width).

§2 explains this method in the one-dimensional case. A simplified version of the iterative process is described in §3, while its final stage is developed in §4. §5 contains those specifications on which the program is based. Numerical results are reported in §6, which also includes a comparison of this method with a direct one.

The basic idea - using auxiliary systems of difference equations corresponding to coarser grids - has been developed independently by the author, but it was already described by R.P.Fedorenko [4,5] in 1961. Since then this idea has only been revived by N. S. Bakhvalov [1] and A. Brandt [2][1]).

§ 2 The one-dimensional case

The boundary value problem for the ordinary differential equation

$$u''(x) = q(x) \quad (0 \leq x \leq 1), \qquad u(0) = u(1) = 0$$

leads to the difference equations

(3) $$S_h u_h = f, \text{ where } S_h = \begin{bmatrix} 2 & -1 & & 0 \\ -1 & 2 & -1 & \\ & \ddots & \ddots & \\ 0 & & -1 & 2 \end{bmatrix}, \quad f = h^2 \begin{bmatrix} q(h) \\ \vdots \\ q(1-h) \end{bmatrix}.$$

Here $h = 1/n$ denotes the step width. n is assumed to be an even number.

The central point of the iterative process is the combination of a *"smoothing procedure"*, annihilating the rapid oscillations, and a *"correction by approximation"*, that especially diminishes the smooth components. Smoothing by Gauß-Seidel iterations (relaxation) suggests itself by its simplicity. Actually, relaxation turns out to be even optimal, since it produces a direct method (cf. [7]), provided that the grid points are arranged in a suitable way. In order to get results which are also typical for

---

1) The author wishes to express his gratitude to Mr. O. Widlund for bringing papers [1, 2] to his attention.

the two-dimensional case, we choose Jacobi iterations damped by the factor $\omega = 1/2$:

(4) $$u \mapsto Gu + \frac{1}{4} f, \quad \text{where} \quad G = I - \frac{1}{4} S_h.$$

By ν-times repeated application of (4) we obtain

$$u \mapsto G^{\nu} u + G_{f,\nu} f, \quad \text{where} \quad G_{f,\nu} = \frac{1}{4} \sum_{\mu=0}^{\nu-1} G^{\mu} .$$

Let $\tilde{u}$ be a value approaching the exact solution u_h of (3). The defect corresponding to $\tilde{u}$ is denoted by

$$d = S_h \tilde{u} - f.$$

The *"correction by approximation"* mentioned above requires the approximate computation of the correction

$$\delta u = \tilde{u} - u_h = S_h^{-1} d .$$

Therefore, we condense $d \in R^{n-1}$ (R: set of real numbers) to the vector

$$\hat{d} = \varphi^T d \in R^{\frac{n}{2}-1}, \quad \text{where} \quad \varphi^T = \begin{bmatrix} 1 & 2 & 1 & & & & \\ & & 1 & 2 & 1 & & \\ & & & & \ddots & & \\ & & & & & 1 & 2 & 1 \end{bmatrix} .$$

The solution v of the equation

$$S_{2h} v = \hat{d}$$

is to be computed in the coarser grid of step size 2h (The one-stage iteration (cf. §3) needs the exact value v, the recursive method (cf. §4) uses an approximation of v). The matrix S_{2h} is defined by (3) with the doubled mesh width 2h, but it can also be obtained by the product $\frac{1}{2}\varphi^T S_h \varphi$. The vector $w = \frac{1}{2} \varphi v \in R^{n-1}$ may be regarded as a continuation of the grid function v to the refined mesh by linear interpolation. Since we expect the vector

$$w = \frac{1}{2} \varphi S_{2h}^{-1} \varphi^T (S_h \tilde{u} - f) = K\tilde{u} - K_f f \qquad (K_f = \frac{1}{2} \varphi S_{2h}^{-1} \varphi^T, \quad K = K_f S_h)$$

to be an approximation of δu, the mapping $\tilde{u} \mapsto \tilde{u} - w$ is called *"correction by approximation"*.

The combination of ν smoothing steps with the correction explained above yields one iteration step:

$$u^{(\mu-1)} \mapsto \tilde{u} = G^{\nu} u^{(\mu-1)} + G_{f,\nu} f \mapsto u^{(\mu)} = (I-K)\tilde{u} + K_f f = M_{\nu}^h u^{(\mu-1)} + M_{f,\nu} f,$$

where $M_{\nu}^h = (I-K)G^{\nu}$. The characteristic properties of convergence are (cf. [6]):

i) The norms of the matrices M_{ν}^h are bounded by

$$\| M_{\nu}^h \|_2 \leq m_{\nu} < 1 \qquad (\nu > 0),$$

where the numbers m_{ν} are independent of h.

ii) The bounds m_{ν} converge to zero as $1/\nu$.

(Values: $m_1 = 0.5$, $m_2 = 0.25$, $m_3 < 0.1502$, $m_{\nu} \approx \sqrt{2}\,(e\nu)^{-1}$).

§ 3 The one-stage method

Coming back to Poisson's equation (1), (2), we define the mesh with a positive step width $\bar{h}$ by

$$G(\bar{h}) = \{(x,y) \in G \mid x/\bar{h} \in Z,\ y/\bar{h} \in Z\},$$

where Z is the set of integers. Since G is assumed to be an open set, $G(\bar{h})$ contains no points of the boundary. We now fix a mesh size h and construct the sequence

$$G(h) \subset G(2h) \subset \dots \subset G(2^i h) \subset \dots \subset G(2^m h) \qquad (o \le i \le m).$$

m is so chosen that $|G(2^m h)|$ is a small number and satisfies

$$o < |G(2^m h)| < |G(2^{m-1}h)|$$

($|G(\bar{h})|$ denotes the number of elements of $G(\bar{h})$).

The grid functions defined on $G(2^i h)$ constitute the vector space V_i ($\dim V_i = |G(2^i h)|$). We discretize (1) and (2) by a system of difference equations:

$$(5) \qquad S_o u_o = f_o \qquad (u_o, f_o \in V_o)$$

(cf. §5). It is assumed that S_o represents a five- or nine-point-formula and that S_o coincides with

$$(6) \qquad \begin{aligned} (S_o u)_{lk} = {} & u_{lk} + \sigma_1 (u_{l+1,k} + u_{l-1,k} + u_{l,k+1} + u_{l,k-1}) \\ & + \sigma_2 (u_{l+1,k+1} + u_{l-1,k+1} + u_{l+1,k-1} + u_{l-1,k-1}) \end{aligned} \qquad \left(o < l,k < h^{-1};\ \sigma_1, \sigma_2 \le o;\ \sigma_1 + \sigma_2 = -\tfrac{1}{4}\right)$$

for the special case of the square $G=(o,1)\times(o,1)$ and $h^{-1} \in Z$. Here we use the following notation: $u_{lk} = u((l,k)\cdot 2^i h)$ is the value of the mesh function $u \in V_i$ at the grid point $(l,k)\cdot 2^i h$ of $G(2^i h)$. For $(l,k)\cdot 2^i h \notin G(2^i h)$ we formally define $u_{lk} := o$.

For the fast iterative solution of the system (5) auxiliary equations

$$(7) \qquad S_i u_i = f_i \qquad (1 \le i \le m;\ u_i, f_i \in V_i)$$

are to be constructed. Therefor we define the linear mappings

$$\begin{aligned} \varphi_i &: V_i \to V_{i-1} \quad \text{(injective)} \\ \psi_i &: V_{i-1} \to V_i \quad \text{(surjective)} \end{aligned} \qquad (1 \le i \le m).$$

It is assumed that φ_i^T and ψ_i are restrictions of nine-point-formulae to the coarser grid $G(2^i h)$. Furthermore

$$\psi_i = \varphi_i^T$$

and

$$(8) \qquad \begin{aligned} (\psi_i u)_{\frac{l}{2},\frac{k}{2}} = {} & u_{lk} + \tfrac{1}{2} (u_{l+1,k} + u_{l-1,k} + u_{l,k+1} + u_{l,k-1}) \\ & + \tfrac{1}{4} (u_{l+1,k+1} + u_{l-1,k+1} + u_{l+1,k-1} + u_{l-1,k-1}) \end{aligned}$$

must be valid for

$$(\frac{l}{2},\frac{k}{2})\cdot 2^i h \in G(2^i h) \subset G(2^{i-1}h),$$

if $G=(o,1)\times(o,1)$ and $(2^i h)^{-1}\in Z$. The equations (6) and (8) should also be fulfilled if the corresponding point of the grid has a sufficiently great distance (proportional to h) from the boundary (cf. [6]).

We define the matrices S_i of system (7) successively by

$$(9) \qquad S_i = \psi_i\, S_{i-1}\, \varphi_i \qquad\qquad (1 \leqslant i \leqslant m),$$

where S_i turns out to be a nine-point-formula, too (cf. [6]). In the papers [1,2,4,5] these matrices are gained by discretizing the differential equation (1) with the enlarged mesh size $2^i h$.

The correction by approximation is given by the mapping

$$(1o) \qquad \begin{bmatrix} u \\ f_i \end{bmatrix} \mapsto \tilde{u} = u - \varphi_{i+1}\, S_{i+1}^{-1}\, \psi_{i+1}\, (S_i u - f_i) = B_i\, u + A_i\, f_i \qquad (o \leqslant i \leqslant m-1),$$

where

$$(11) \qquad \begin{aligned} A_i &= \varphi_{i+1}\, S_{i+1}^{-1}\, \psi_{i+1}\,, \\ B_i &= I - A_i\, S_i\,. \end{aligned}$$

From (9) it may be concluded that $A_i S_i$ and B_i are projections. §5 will describe in detail two smoothing procedures

$$\begin{bmatrix} u \\ f_i \end{bmatrix} \mapsto C_i^{I} \begin{bmatrix} u \\ f_i \end{bmatrix}, \qquad \begin{bmatrix} u \\ f_i \end{bmatrix} \mapsto C_i^{II} \begin{bmatrix} u \\ f_i \end{bmatrix},$$

where the matrices are splitted into the four blocks:

$$C_i^{I} = \begin{bmatrix} C_{u,i}^{I} & C_{f,i}^{I} \\ 0 & I \end{bmatrix} : V_i^{\,2} \longrightarrow V_i^{\,2} \quad \text{and} \quad C_i^{II} = \begin{bmatrix} C_{u,i}^{II} & C_{f,i}^{II} \\ 0 & I \end{bmatrix}.$$

Putting all together, one iteration step

$$(12) \qquad \begin{bmatrix} u_i^{(\mu)} \\ f_i \end{bmatrix} = J_i \begin{bmatrix} u_i^{(\mu-1)} \\ f_i \end{bmatrix} \qquad\qquad (o \leqslant i \leqslant m-1)$$

is obtained, where

$$J_i = \begin{bmatrix} J_{u,i} & J_{f,i} \\ 0 & I \end{bmatrix} = C_i^{II} \begin{bmatrix} B_i & A_i \\ 0 & I \end{bmatrix} C_i^{I}\,.$$

In the case of a rectangle G it is not difficult to prove (cf. [6]), that the rate of convergence has properties corresponding to i) and ii) of §2. Especially for any $\kappa \in (o,1)$ the smoothing procedures C_i^{I} and C_i^{II} $(o \leqslant i \leqslant m-1)$ can be so chosen, that $\|J_{u,i}\|_2$ is bounded by κ for any step width $2^i h$.

§ 4 The recursive method

The correcting step (1o) uses the exact solution u_{i+1} of $S_{i+1}u_{i+1} = f_{i+1} := \psi_{i+1}(S_i u - f_i)$. Substituting u_{i+1} by an approximation we obtain the recursive method. One step of the new iteration has the representation (12) with J_i replaced by

$$\tilde{J}_i = \begin{bmatrix} \tilde{J}_{u,i} & \tilde{J}_{f,i} \\ 0 & I \end{bmatrix} = C_i^{II} \begin{bmatrix} \tilde{B}_i & \tilde{A}_i \\ O & I \end{bmatrix} C_i^{I} \qquad (0 \leq i \leq m-1),$$

where $\tilde{B}_i = I - \tilde{A}_i S_i$. The matrices $\tilde{A}_i$ will be determined recursively for $i=m-1,m-2,\ldots,0$. There are several possibilities for defining $\tilde{A}_{m-1}$. Since, by choice of m, $|G(2^m h)|$ is a small number, equation (7) can easily be resolved for $i=m$. Thus, we may choose

$$\tilde{A}_{m-1} = A_{m-1} = \varphi_m S_m^{-1} \psi_m.$$

The other matrices $\tilde{A}_i$ are given by

$$(13) \qquad \tilde{A}_i = \varphi_{i+1} [I,0] \tilde{J}_{i+1}^{\gamma} \begin{bmatrix} 0 \\ I \end{bmatrix} \psi_{i+1} = \varphi_{i+1} \left(\sum_{\nu=0}^{\gamma-1} \tilde{J}_{u,i+1}^{\nu} \right) \tilde{J}_{f,i+1} \psi_{i+1} \qquad (0 \leq i < m-1),$$

where γ is the number of the secondary iterations. A comparison with (11) shows, that $S_{i+1}^{-1} f_{i+1}$ is replaced by an approximation, which is received from the starting value $u = 0$ by γ iterations involving $\tilde{J}_{i+1}$.

The program of §6 uses γ smoothing steps at the level $i=m$ instead of inverting S_m. Formally $\tilde{A}_{m-1}$ can be obtained from (13) by setting $\tilde{J}_m := C_m^{II} C_m^{I}$.

The following Note reduces the convergence properties of the recursive process to those of the one-stage method of §3:

Note 1: We assume that there are constants A, B ≥ 0 and $\kappa \in (0,1)$ (independent of h) such that

$$(14a) \qquad \| C_{u,i}^{II} \varphi_{i+1} \|_2 \leq A \qquad (\text{resp. } \| S_i C_{u,i}^{II} \varphi_{i+1} S_{i+1}^{-1} \|_2 \leq A),$$

$$(14b) \qquad \| S_{i+1}^{-1} \psi_{i+1} S_i C_{u,i}^{I} \|_2 \leq B \qquad (\text{resp. } \| \psi_{i+1} S_i C_{u,i}^{I} S_i^{-1} \|_2 \leq B),$$

$$(14c) \qquad \| J_{u,i} \|_2 \leq \kappa, \ \| \tilde{J}_{u,m-1} \|_2 \leq \kappa \qquad (\text{resp. } \| S_i J_{u,i} S_i^{-1} \|_2 \leq \kappa, \ \| S_{m-1} \tilde{J}_{u,m-1} S_{m-1}^{-1} \|_2 \leq \kappa).$$

Supposing the existence of a number $\tilde{\kappa} > 0$ with

$$(14d) \qquad \kappa + AB \tilde{\kappa}^{\gamma} \leq \tilde{\kappa} < 1,$$

we can bound the convergence rate of the recursive method by $\tilde{\kappa}$:

$$(15) \qquad \| \tilde{J}_{u,i} \|_2 \leq \tilde{\kappa} < 1 \qquad (\text{resp. } \| S_i \tilde{J}_{u,i} S_i^{-1} \|_2 \leq \tilde{\kappa} < 1).$$

For $\gamma = 1$ the inequality (14d) leads to

$$AB < 1, \quad \tilde{\kappa} := \kappa/(1-AB) < 1.$$

For $\gamma = 2$, (14d) is equivalent to

$$4\kappa AB \leq 1, \quad \tilde{\kappa} := 2\kappa/(1 + \sqrt{1-4\kappa AB}) < 1.$$

Let $C\{v \mapsto w\}$ denote the number of operations $(+,-,\times,/)$ necessary for performing the mapping $v \mapsto w$. The next Note shows that the amount of computational work of one overhead iteration (i.e. one iteration step at level $i = o$) is proportional to h^{-2}:

Note 2 (operation count): If α_1, α_2, α_3 and α_4 are constants with

$$C\{(u,f) \mapsto C^{I}_{u,i}u + C^{I}_{f,i}f\} \le \alpha_1 (2^i h)^{-2} \qquad \text{(smoothing by } C^{I}_{i}\text{)},$$

$$C\{(u,f) \mapsto C^{II}_{u,i}u + C^{II}_{f,i}f\} \le \alpha_2 (2^i h)^{-2} \qquad \text{(smoothing by } C^{II}_{i}\text{)},$$

$$C\{(u,f) \mapsto \psi_{i+1}(S_i u - f)\} \le \alpha_3 (2^i h)^{-2} \qquad \text{(evaluation of the defect)},$$

$$C\{(u,f) \mapsto u - \varphi_{i+1}\, v\} \le \alpha_4 (2^i h)^{-2} \qquad \text{(correction)},$$

$$C\{(u,f) \mapsto \tilde{J}_{u,m-1}u + \tilde{J}_{f,m-1}f\} \le \sum_{j=1}^{4} \alpha_j \,(2^{m-1}h)^{-2}$$

for $o \le i \le m-1$ and if $1 \le \gamma \le 3$, then one iteration requires

$$C\left\{\begin{bmatrix} u^{(\mu-1)} \\ f \end{bmatrix} \mapsto \begin{bmatrix} u^{(\mu)} \\ f \end{bmatrix} = \tilde{J}_0 \begin{bmatrix} u^{(\mu-1)} \\ f \end{bmatrix}\right\} < \frac{4}{4-\gamma}\left(\sum_{j=1}^{4} \alpha_j\right) h^{-2}$$

operations.

Because of (15), an accuracy ε is attainable by $O(|\log \varepsilon|)$ iteration steps. Thus, the total amount is of order $O(h^{-2}|\log \varepsilon|)$.

§ 5 *Specifications*

Sections §3 and §4 give the framework of the presented method. Here we discuss the the details leading to the program on which the numerical results of §6 are based.

5.1 Choice of S_o:

The discretization of (1), (2) in a general region by a five-point-formula is well-known. It is first described by Shortley and Weller [8]. Extending this method to nine-point-formulae, we get the matrix S_o with the property (6) (cf. [6]). S_o depends on the parameters σ_1 and σ_2 $(\sigma_1 \le o,\ \sigma_2 \le o,\ \sigma_1+\sigma_2=-1/4)$. The choice $\sigma_1=-1/4$ and $\sigma_2=o$ yields the five-point-formula mentioned above. For $\sigma_1=-1/5$ and $\sigma_2=-1/2o$ we obtain the matrix S_o of the Mehrstellenverfahren (cf. [3]).

5.2 Choice of φ_i and ψ_i:

For $u \in V_o$ and $(ih,jh) \in G(2h) \subset G(h)$ we set

$$(16) \qquad (\psi_1 u)(ih,jh) = \sum_{l=i-1}^{i+1} \sum_{k=j-1}^{j+1} p^{lk}_{ij}\, u_{lk}\,.$$

The coefficients are defined by $p^{lk}_{ij} = o$, if $(lh,kh) \notin G(h)$; otherwise

$$(17a) \qquad p^{ij}_{ij} = 1,$$

$$(17b) \qquad p^{i+1,j}_{ij} = \frac{d-h}{d}, \quad \text{where } d = \min(2h,\ \sup\{\xi > o \mid (ih+\theta,jh) \in G \text{ for } \theta \in [o,\xi]\}).$$

$p^{i-1,j}_{ij}$ and $p^{i,j\pm 1}_{ij}$ are determined similarly to (17b), whereas the coefficients $p^{i\pm 1,j\pm 1}_{ij}$

are so chosen that

(18) $\quad (S_o \psi_1^T \delta_{ih,jh}) ((i\pm1)h, (j\pm1)h) = o,$

where $\delta_{ih,jh} \in V_1$ takes the value 1 at the point (ih,jh) and vanishes at the remaining part of G(2h).

For i>1 the mappings ψ_i are defined in a manner analogous to (16),(17),(18) (cf. [6]). φ_i results from ψ_i by transposition:

(19) $\quad \varphi_i = \psi_i^T \qquad (1 \leq i \leq m) .$

5.3 Choice of C_i^I, C_i^{II} and γ:

Note 2 points out that 1, 2 or 3 are the only suitable values for γ. Numerical experiences show that the transition form γ=2 to γ=3 does not improve the results. γ=2 produces better rates of convergence but it requires more computational work than γ=1. In all we prefer γ=1.

Defining

$$G_{ij} = \{(x,y) \in G(2^i h) \mid (x,y)=(l,k)\cdot 2^i h,\ l \text{ is } \begin{Bmatrix}\text{even } (j=o,2)\\ \text{odd } (j=1,3)\end{Bmatrix}, k \text{ is } \begin{Bmatrix}\text{even } (j=o,1)\\ \text{odd } (j=2,3)\end{Bmatrix} \},$$

the mesh $G(2^i h)$ is the union of the distinct sets G_{ij} (j=o,1,2,3). The matrix R_{ij} ($o \leq i \leq m$, $o \leq j \leq 3$) describes the resolving of those equations of (5) or (7) respectively, that are associated with points of the subset G_{ij} ; i.e.

$$\begin{bmatrix}\tilde{u}\\ f_i\end{bmatrix} = R_{ij}\begin{bmatrix}u\\ f_i\end{bmatrix} \Leftrightarrow (\tilde{u}-u)\Big|_{G_{ik}} = o \text{ for } k\neq j \text{ and } (S_i\tilde{u} - f_i)\Big|_{G_{ij}} = o.$$

The program uses the smoothing procedures

$$C_i^I = R_{i3}$$

$$C_i^{II} = R_{i3}C_i \text{ for } i>o; \quad C_o^{II} = C_o \text{ for } i=o, \text{ where } C_i = R_{io}R_{i2}R_{i1}R_{io}R_{i3}R_{i2}R_{i1}\,.$$

C_i^I and C_i^{II} correspond nearly to two Gauß-Seidel iterations.

5.4 Further details:

All results of §6 are obtained with the starting value u = o. The iterations are proceded by one smoothing step at the level i=o.

The present program requires less than

(2o) $\quad 41\, h^{-2}\, F(G) + L(\partial G)\, O(h^{-1})$

operations per iteration. Herein F(G) and L(∂G) denote the area of the region G and the arc length of its boundary respectively. In case of σ_2=o (i.e. S_o corresponds to the five-point-formula) it is possible to reduce the number 41 of (2o) to 27.6 .

The matrices φ_i, ψ_i and S_i must be computed once and for all in a preprocessing phase. The respective number of operations is of order $L(\partial G)\cdot O(h^{-1})$, since values deviating from (6) and (8) only arise in the neighbourhood of the boundary ∂G.

§ 6 Numerical results

The numerical computations are performed on the CDC-computer Cyber 76 of the Rechenzentrum der Universität zu Köln. The FORTRAN compiler FTN 4.2 (OPT=2) has been used.

6.1 Results of the recursive method:

In the following we solve the boundary value problem (1), (2) with

$$q(x,y) = -4, \quad r(x,y) = x^2 + y^2$$

for different regions $G \subset [0,1]\times[0,1]$. Thus, $u_0 = x^2 + y^2$ is the exact solution of the differential equation (1) and of the difference equation (5).

Table 1 shows the convergence of the recursive method for the square $G = G_1 := (0,1)\times(0,1)$ covered by a mesh of size $h = 1/64$. The Euclidean norms and the maximum norms of the differences $u^{(\mu)} - u_0$ are listed for both discretizations (five-point-formula and Mehrstellenverfahren, cf. §5.1). It is recalled that $u^{(1)}$ is obtained from $u^{(0)} = 0$ by one smoothing step. The quotients of the successive differences are placed beside these columns. For this and for most of the other examples the differences become smaller than machine precision before the quotients approach a limit. Therefore, we shall not compare the true rates of convergence but the mean values

$$\left(\| u^{(\mu)} - u_0 \|_2 \,/\, \| u^{(3)} - u_0 \|_2 \right)^{1/(\mu-3)},$$

where μ is the greatest integer such that $u^{(\mu)}$ is not impaired by the influence of rounding errors.

The averaged convergence rates for different mesh sizes are listed in Table 2. They confirm that the speed of convergence is far away from approaching the value one. It is typical that the Mehrstellenverfahren converges faster than the five-point-formula. On the other hand the five-point-formula requires less computational work, although the used program is not optimal for this discretization.

	five-point-formula				Mehrstellenverfahren			
μ	$\|u^{(\mu)}-u_0\|_2$		$\|u^{(\mu)}-u_0\|_\infty$		$\|u^{(\mu)}-u_0\|_2$		$\|u^{(\mu)}-u_0\|_\infty$	
1	$7.43_{10}-1$		$1.82_{10}+0$		$7.37_{10}-1$		$1.81_{10}+0$	
2	$2.48_{10}-2$	0.0333	$6.38_{10}-2$	0.0351	$2.28_{10}-2$	0.0309	$5.48_{10}-2$	0.0303
3	$9.95_{10}-4$	0.0402	$3.67_{10}-3$	0.0575	$7.31_{10}-4$	0.0321	$2.84_{10}-3$	0.0518
4	$5.51_{10}-5$	0.0554	$2.10_{10}-4$	0.0572	$3.06_{10}-5$	0.0418	$1.18_{10}-4$	0.0415
5	$2.99_{10}-6$	0.0542	$1.16_{10}-5$	0.0552	$1.11_{10}-6$	0.0363	$4.25_{10}-6$	0.0360
6	$1.61_{10}-7$	0.0538	$6.38_{10}-7$	0.0550	$3.62_{10}-8$	0.0326	$1.40_{10}-7$	0.0329
7	$8.54_{10}-9$	0.0530	$3.43_{10}-8$	0.0538	$1.12_{10}-9$	0.0309	$4.21_{10}-9$	0.0301
8	$4.50_{10}-10$	0.0527	$1.84_{10}-9$	0.0536	$3.63_{10}-10$	0.0324	$1.32_{10}-10$	0.0314
9	$2.36_{10}-11$	0.0524	$9.78_{10}-11$	0.0532				
10	$1.25_{10}-12$	0.0530	$5.17_{10}-12$	0.0529				

Table 1: Results for the square G_1, h=1/64

inverse mesh width 1/h	m (cf.§3)	five-point-formula			Mehrstellenverfahren		
		averaged rate of conv.	CPU-time (sec) for prepro-cessing	one iteration	averaged rate of conv.	CPU-time (sec) for prepro-cessing	one iteration
8	2	o.o392	o.oo9	o.oo3	o.o122	o.oo9	o.oo3
12	2	o.o488	o.o19	o.oo6	o.o197	o.o2o	o.oo6
16	3	o.o489	o.o36	o.o12	o.o196	o.o37	o.o12
24	3	o.o487	o.o65	o.o22	o.o184	o.o66	o.o23
32	4	o.o5o1	o.122	o.o51	o.o222	o.122	o.o53
48	4	o.o514	o.199	o.o94	o.o266	o.2oo	o.o95
64	5	o.o535	o.338	o.2o3	o.o346	o.339	o.212
96	5	o.o561	o.58o	o.428	o.o428	o.58o	o.447
128	6	o.o576	o.917	o.799	o.o47o	o.919	o.834

Table 2: Comparison of different mesh sizes for the square G_1

Comparing the data for the preprocessing phase (construction of φ_i and S_i for i>o), we see that this part of computational work is of some order between $O(h^{-1})$ and $O(h^{-2})$, while the amount of one iteration increases like $O(h^{-2})$ according to Note 2. For the sake of completeness it should be mentioned that the time for computing $u^{(1)}$ is about 4o% of the run of one iteration.

For the case of the square G_1 discussed above, the mappings φ_i and ψ_i (i>o) do not depend on the construction (16) and (19), since their coefficients coincide with the regular values (8). Nevertheless, it may be stated for a large family of irregular regions that the convergence rate does not change for the worse.

The boundary value problems are solved below for the regions G_i ($1 \le i \le 12$) sketched in Table 3. The square G_1, the circle G_2 and the ellipse G_3 are convex regions. The example G_4 (G_5) is a square (circle) with one circle of radius 1/4 (four circles of radii 1 / 1o) cut out. G_6 is bounded by four arcs forming a zero-degree angle at the edges. The remaining examples are regions with boundaries containing an angle of $3\pi/2$, $7\pi/2$ and 2π respectively.

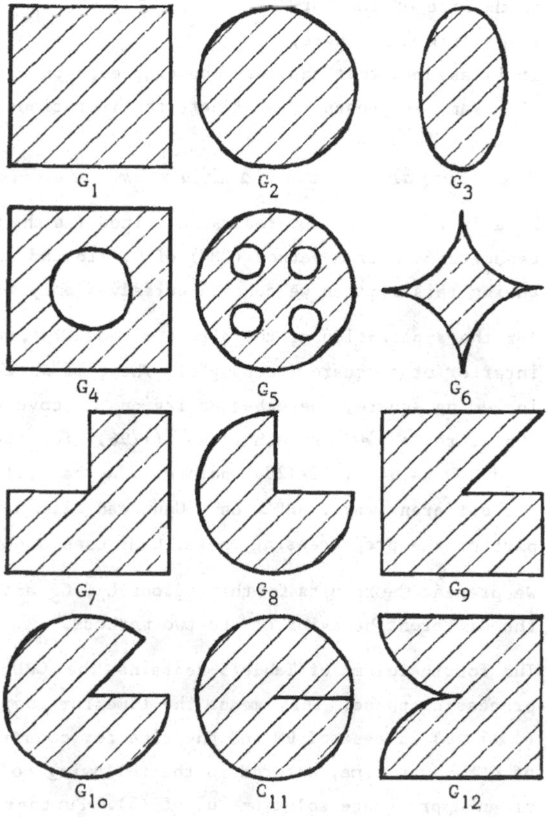

Table 3: The different regions G_i

G	averaged rates of convergence		$\|u^{(5)}-u_o\|_2$ $(h=\frac{1}{64})$	
	5-point formula	Mehrstellen-verfahren	5-point formula	Mehrst. verf.
G_1	o.o5o - o.o56	o.o22 - o.o43	3.o$_{1o}$-6	1.1$_{1o}$-6
G_2	o.o31 - o.o45	o.o23 - o.o25	2.4$_{1o}$-7	1.7$_{1o}$-7
G_3	o.o4o - o.o43	o.o15 - o.o21	2.3$_{1o}$-7	4.9$_{1o}$-8
G_4	o.o49 - o.o55	o.o28 - o.o42	4.2$_{1o}$-6	2.3$_{1o}$-6
G_5	o.o45 - o.o58	o.o34 - o.o44	5.5$_{1o}$-6	3.3$_{1o}$-6
G_6	o.o33 - o.o4o	o.o18 - o.o29	4.7$_{1o}$-8	2.4$_{1o}$-8
G_7	o.o85 - o.o96	o.o66 - o.o83	2.1$_{1o}$-5	1.4$_{1o}$-5
G_8	o.o84 - o.o95	o.o63 - o.o84	2.3$_{1o}$-5	1.4$_{1o}$-5
G_9	o.1o6 - o.125	o.o91 - o.117	6.o$_{1o}$-5	4.5$_{1o}$-5
G_{1o}	o.1o5 - o.127	o.o88 - o.119	5.o$_{1o}$-5	3.8$_{1o}$-5
G_{11}	o.124 - o.158	o.11o - o.152	1.4$_{1o}$-4	1.2$_{1o}$-4
G_{12}	o.12o - o.154	o.1o8 - o.147	1.2$_{1o}$-4	1.o$_{1o}$-4

Table 4: Intervals of the rates of convergence for the different regions

Table 4 contains the averaged rates of convergence for the mesh widths h = 1/32, 1/48, 1/64 and 1/96. Comparing these rates for the different regions G_i, we note from Table 4, that all regions give results similar to those of the square G_1, if their boundaries contain no angle greater than 18o°. Otherwise, the rate of convergence depends only on the size of the corresponding angle. Here, it is assumed that the corresponding edge is a point of the grid as was the case for all examples presented (if not, the iterations converge somewhat faster).

6.2 Comparison with a direct method (capacitance matrix method):

Finally we compare our recursive method (in the following abbreviated by RM) with the capacitance matrix method (CMM) of O. Widlund and W. Proskurowski, described in [9]. During this section we concern ourselves only with the five-point-formula.

For the application of the direct method CMM, the region G must be embedded in the interior of a square (rectangle). Thus, if we choose a grid of 64×64 points for the including square, the embedded region is covered with less than 48×48 grid points. Therefore, a mesh size $h \geq 1/48$ (1/96) for the regions G_i of §6.1 corresponds to a grid of 64×64 (128×128) points. The capacitance matrix equation can be solved in two different ways. CMM2 uses Gaussean elimination. The factoring of the matrix is part of the preprocessing phase. CMM1 uses a conjugate gradient method.

We present the results for the regions G_2, G_5 and G_{11}, since these examples illustrate the different behavior of the two methods.

The fourth column of Table 5 contains the CPU-time (in seconds) needed for the preprocessing phase. This means the time for computing the coefficients of φ_i and S_i $(1 \leq i \leq m)$ in case of RM and the time for constructing the capacitance matrix in case of CMM. The time, entered in the following column, is necessary for the computation of an approximate solution $\tilde{u}$ of (5). Further the norms of the differences $\tilde{u}-u_0$ $(u_0 = S_0^{-1} f_0)$ and the number i of iterations are given in the table. For RM, i is the

region	$\frac{1}{h}$	program	preprocessing phase	time for computing $\tilde{u}$	accuracy $\lVert\tilde{u}-u_0\rVert_q=\varepsilon_q$ ε_2	ε_∞	iterations	storage requirement
G_2	24	CMM1	0.17	0.21	$1.8_{10}{-8}$	$5.7_{10}{-8}$	13	7680
:	:	CMM2	0.41	0.09	$1.7_{10}{-12}$	$3.2_{10}{-12}$	-	:
:	48	RM	0.21	0.45	$9.5_{10}{-9}$	$4.6_{10}{-8}$	6	9948
:	48	CMM1	0.72	0.89	$3.0_{10}{-7}$	$3.9_{10}{-6}$	12	28780
:	:	CMM1	:	1.02	$5.1_{10}{-9}$	$4.2_{10}{-8}$	16	:
:	:	CMM1	:	1.12	$4.7_{10}{-11}$	$5.1_{10}{-10}$	19	:
:	:	CMM2	2.87	0.42	$5.7_{10}{-12}$	$1.1_{10}{-11}$	-	:
:	64	RM	0.34	1.11	$3.5_{10}{-10}$	$2.1_{10}{-9}$	7	16242
:	86	CMM1	2.51	3.96	$3.7_{10}{-9}$	$4.1_{10}{-8}$	17	96144
:	:	CMM2	15.57	1.89	$1.2_{10}{-11}$	$2.6_{10}{-11}$	-	:
:	96	RM	0.58	2.31	$6.1_{10}{-10}$	$3.9_{10}{-9}$	7	32862
:	128	RM	0.89	2.27	$1.7_{10}{-5}$	$8.7_{10}{-5}$	4	55020
:	:	RM	:	4.25	$8.9_{10}{-10}$	$5.1_{10}{-9}$	7	:
G_5	24	CMM1	0.56	1.00	$1.6_{10}{-6}$	$5.9_{10}{-6}$	28	20340
:	:	CMM2	2.34	0.10	$3.0_{10}{-12}$	$4.7_{10}{-12}$	-	:
:	48	RM	0.37	0.47	$5.8_{10}{-8}$	$2.3_{10}{-7}$	6	13766
:	48	CMM1	2.21	3.27	$1.1_{10}{-6}$	$4.0_{10}{-6}$	23	73468
:	:	CMM2	15.94	0.47	$3.9_{10}{-12}$	$7.8_{10}{-12}$	-	:
:	64	RM	0.58	0.93	$3.2_{10}{-7}$	$1.2_{10}{-6}$	6	21358
G_{11}	48	RM	0.25	0.72	$1.9_{10}{-8}$	$5.5_{10}{-8}$	9	11212
:	48	CMM1	1.23	3.78	$5.9_{10}{-9}$	$6.3_{10}{-8}$	54	44098
:	:	CMM2	6.54	0.44	$5.6_{10}{-12}$	$1.1_{10}{-11}$	-	:
:	64	RM	0.41	1.50	$6.9_{10}{-8}$	$1.9_{10}{-7}$	9	17896

Table 5: Comparison of RM and CMM

integer with $u^{(i)}=\tilde{u}$, while for CMM1 the solution of the capacitance matrix equation is approximated by i iteration steps (In CMM1 the number i and the accuracy of $\tilde{u}$ is controlled by a given tolerance). The determination of the grid G(h) and the construction of the matrix S_0 and the vector f_0 are not taken into account.

Obviously, RM is faster than CMM1. It is advisable to use CMM2 only, if the work for the preprocessing phase is irrelevant. Even in this case the program RM needs a CPU-time similar to the time of CMM2, if one is satisfied by the accuracy attained by six iterations (cf. also Table 1 and 4).

In addition Table 5 reveals an essential disadvantage of capacitance matrix methods: The magnitude, determining the computational work, is the number p of irregular points (having neighbours on the boundary). The integers s - number of equations of the system (5) - and t - number of grid points of the embedding rectangle - are of secondary account. The CPU-time for the preprocessing phase is at least of order $O(p^2)$ for the capacitance matrix methods. The corresponding amount of RM is composed by $O(p)$ operations - involved by the evaluation of (9) and (17) for each irregular point - and by s checks, testing if the repective point is regular or not. The computational work of the main part behaves like $O(s)+O(p)$ for RM and like $O(p^2)$ for CMM. The values of the fifth column in Table 5 indicate this dependence, because p increases, t remains constant and s decreases, if we change from G_2 to G_5. The storage requirement amount to $p^2+2t+O(p)+O(\sqrt{t})$ for CMM and $8t/3+O(p)$ for RM. This causes a severe restriction of the mesh width in case of CMM.

We conclude the comparison with a remark about the applicability of RM and CMM. The program CMM requires a special property of the grid G(h): All points of G(h) must have at most one neighbour on the boundary in the vertical direction and at most one in the horizontal direction. Thus, regions like G_6, G_9 and G_{10} are excluded. Even the circle G_2 do not satisfy this condition for each mesh size $h \in (o,h_o]$. Therefore, it is important to note that the program RM is applicable without exception.

References:

1. Bakhvalov, N.S.: On the convergence of a relaxation method with natural constraints on the elliptic operator, U.S.S.R. Computational Mathematics and Mathematical Physics, vol. 6, pp. 1o1 - 135, 1966 (Zh. vÿchisl. Mat. mat. Fiz., 6, 5, 861-885, 1966)
2. Brandt, A.: Multi-level adaptive technique (MLAT) for fast numerical solution to boundary value problems, Proceedings of the Third International Conference on Numerical Methods in Fluid Mechanics, 1972, pp. 82 - 89, ed. by H. Cabannes and R. Temam, Lecture Notes in Physics, 18, Springer-Verlag, Berlin, Heidelberg, New York, 1973
3. Collatz, L.: The numerical treatment of differential equations, 3rd ed., Springer-Verlag, Berlin, Heidelberg, New York, 1966
4. Fedorenko, R.P.: A relaxation method for solving elliptic difference equations, U.S.S.R. Computational Mathematics and Mathematical Physics, vol. 1, pp. 1o92-1o96, 1962 (Zh. vÿchisl. Mat. mat. Fiz., 1, 5, 922 - 927, 1961)
5. Fedorenko, R.P.: The speed of convergence of one iterative process, U.S.S.R. Computational Mathematics and Mathematical Physics, vol. 4, pp. 227 - 235, 1964 (Zh. vÿchisl. Mat. mat. Fiz., 4, 3, 559 - 564, 1964)
6. Hackbusch, W.: Ein iteratives Verfahren zur schnellen Auflösung elliptischer Randwertprobleme, Bericht des Mathematischen Instituts der Universität zu Köln - Angewandte Mathematik, 1976, to be published
7. Schröder, J., U. Trottenberg and K. Witsch: On a fast Poisson solver and various applications, this volume
8. Shortley, G.H. and R. Weller: Numerical solution of Laplace's equation, Journal of Applied Physics, vol. 9, pp. 334 - 348, 1938
9. Widlund, O. and W. Proskurowski: On the numerical solution of Helmholtz's equation by the capacitance matrix method, ERDA Research and Development Report, Courant Institut of Mathematical Science, New York University, Nov. 1975

ON THE STABILITY REGIONS OF MULTISTEP MULTIDERIVATIVE METHODS

Rolf Jeltsch

1. Introduction.

Consider the initial value problem

(1) $$y' = f(x,y) \quad , \quad y(0) = \eta \ .$$

The exact solution of (1) is denoted by $y(x)$. Let $h > 0$ be a constant and $x_n = nh$, $n = 0,1,2,\ldots$. A general linear multistep multiderivative method then has the form

(2) $$\sum_{i=0}^{k} \alpha_{i0}\, y_{n+i} + \sum_{j=1}^{\ell} h^j \sum_{i=0}^{k} \alpha_{ij}\, f_{n+i}^{(j-1)} = 0 \quad , \quad n = 0,1,2,\ldots \ ,$$

where $f_{n+i}^{(j)} = f^{(j)}(x_{n+i}, y_{n+i})$ and $f^{(j)}(x,y)$ is the j-th total derivative of $f(x,y(x))$ with respect to x . α_{ij} are real constants and we shall always assume that $\alpha_{k0} \neq 0$, $\sum_{j=0}^{\ell} |\alpha_{0j}| \neq 0$, $\sum_{i=0}^{k} |\alpha_{i\ell}| \neq 0$. We say that (2) defines a (k,ℓ)-method. The starting values $y_0, y_1, \ldots, y_{k-1}$ have to be computed using some other method such as Runge-Kutta. Once these values are known (2) is used recursively to compute y_n , $n = k,k+1,\ldots$. The hope is that y_n is a good approximation to $y(x_n)$.

Let us consider the scalar test initial value problem

(3) $$y' = \lambda y \quad , \quad y(0) = 1 \quad , \quad x \in [0,\infty) \quad ,$$

where λ is an arbitrary complex number. When we apply the (k,ℓ)-method (2) to (3) we obtain the recurrence relation

(4) $$\sum_{i=0}^{k} \sum_{j=0}^{\ell} \alpha_{ij} (\lambda h)^j \, y_{n+i} = 0 \quad , \quad n = 0,1,\ldots \ .$$

This is a linear homogeneous difference equation with constant coefficients with the characteristic equation

(5) $$\Phi(\zeta,\mu) = \sum_{i=0}^{k} \sum_{j=0}^{\ell} \alpha_{ij}\, \mu^j \zeta^i = 0 \quad , \quad \mu := h\lambda \ .$$

Every solution of (4) can be written in the form

(6) $$y_n = \sum_{i=1}^{k} \pi_i(n)\, \zeta_i^n(\mu) \quad ,$$

where $\zeta_i(\mu)$ are the distinct roots of the polynomial $\Phi(\zeta,\mu)$ and $\pi_i(x)$ is a polynomial of degree less than the multiplicity of the root $\zeta_i(\mu)$. Note that for at most ℓ values of μ $\Phi(\zeta,\mu)$ is a polynomial in ζ of degree less than k

and hence the recurrence relation (4) may not have a solution for some starting values $y_0,\dots,y_{k-1}$. (5) represents an algebraic equation. Let $\zeta(\mu)$ denote the algebraic function with

(7) $$\Phi(\zeta(\mu),\mu) \equiv 0$$

and then $\zeta_i(\mu)$ can be interpreted as its branches. Clearly $\zeta_i(0)$ are the roots of

(8) $$\rho_0(\zeta) = \sum_{i=0}^{k} \alpha_{i0}\zeta^i .$$

A method is said to be stable if $\rho_0(\zeta)$ has no root outside the unit circle and the roots of modulus 1 are simple. The (k,ℓ)-method is convergent in the sense of Dahlquist and Henrici if and only if it is stable and there exists a branch $\zeta_1(\mu)$ of $\zeta(\mu)$ which is analytic in a neighborhood of $\mu = 0$ and satisfies

(9) $$\zeta_1(\mu) - e^{\mu} = O(\mu^2) \qquad \text{as} \quad \mu \to 0 .$$

This branch is called principle branch. By analytic continuation we extend it in the largest possible starregion $\Omega \subset \mathbb{C}$, see Fig. 1.

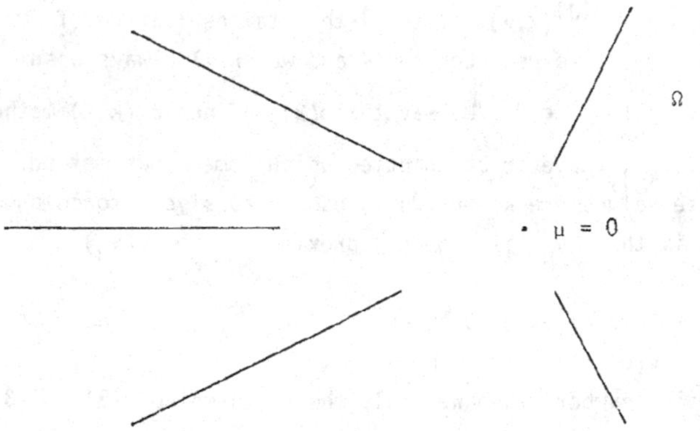

Fig. 1. Region Ω in which $\zeta_1(\mu)$ can be defined.

The method has error order p if and only if

(10) $$\zeta_1(\mu) - e^{\mu} = - c_{p+1}\, \mu^{p+1} + O(\mu^{p+2}) ,$$

where $c_{p+1} \neq 0$. c_{p+1} is called the error constant. This definition for the error order is for convergent methods equivalent to the ones given in Genin [8] , Reimer [16] and Stetter [19] . If we judge the performance of a convergent (k,ℓ)-method only by its error order p and its error constant, then we can without loss of generality assume that $\Phi(\zeta,\mu)$ is irreducible. For consider a method of order p with a certain c_{p+1} for which $\Phi(\zeta,\mu)$ is reducible. Hence $\Phi(\zeta,\mu) = \prod_{s=1}^{t} \Phi_s(\zeta,\mu)$, where all $\Phi_s(\zeta,\mu)$ are irreducible. Let $\zeta_1(\mu)$ be the principle branch of $\zeta(\mu)$ defined by (7) then there exists exactly one s, say q, for which $\Phi_q(\zeta_1(\mu),\mu) \equiv 0$ in a neighborhood of $\mu = 0$, since otherwise $\zeta_1(0)$ would be a multiple root of $\rho_0(\zeta)$. Hence the method given by the coefficients of $\Phi_q(\zeta,\mu)$ is stable, has error order p and error constant c_{p+1}. In the following we shall always assume that the method has an irreducible $\Phi(\zeta,\mu)$ unless we state explicitely the opposite.

The exact solution of (3) is $y(x_n) = e^{\lambda x_n} = (e^{\mu})^n$. Hence from (9) and (6) one finds that the accuracy of the method stems from the principle branch. Hence one is interested in the values of μ for which $\zeta_1(\mu)$ dominates the other branches, i.e. we are interested in the following sets:

$$R_s = \{\mu \in \Omega \mid \ |\zeta_1(\mu)| > |\zeta_i(\mu)| \ , \ i = 2,3,\dots,k\}$$
$$R_e = \{\mu \in \Omega \mid \ |\zeta_1(\mu)| \geq |\zeta_i(\mu)| \ , \ i = 2,3,\dots,k\}$$
$$R_{es} = \{\mu \in \Omega \mid \ |\zeta_1(\mu)| \geq |\zeta_i(\mu)| \ , \ i = 2,3,\dots,k \text{ , and if } = \text{ holds then } \zeta_i(\mu) \text{ is simple}\} .$$

In the literature all these sets are usually referred to as <u>regions</u> of <u>relative stability</u>.

Let us now turn to the problem of absolute stability. Let h be fixed and we let n tend to infinity. Then the solution $y(x_n)$ of (3) tends to 0 if and only if $\text{Re}\,\lambda < 0$ and it tends to ∞ if and only if $\text{Re}\,\lambda > 0$. We are interested for which values of $\mu = h\lambda$ does the numerical solution tend to 0. Hence we consider the sets

$$A_s = \{\mu \in \overline{\mathbb{C}} \mid |\zeta_i(\mu)| < 1 \ , \ i = 1,2,\dots,k\}$$
$$A_e = \{\mu \in \overline{\mathbb{C}} \mid |\zeta_i(\mu)| \leq 1 \ , \ i = 1,2,\dots,k\}$$
$$A_{es} = \{\mu \in \overline{\mathbb{C}} \mid |\zeta_i(\mu)| \leq 1 \ , \ i = 1,2,\dots,k \text{ and if } = \text{ holds then } \zeta_i(\mu) \text{ is simple}\} .$$

In the literature all three sets are usually referred to as <u>regions</u> of <u>absolute stability</u>.

In the next section we shall give some elementary properties of the regions of relative and absolute stability. Further we define symmetric, asymptotically exact and globally asymptotically exact methods, see [9], [14]. In section 3 we shall employ the symmetry priniple by Schwarz to characterize the above mentioned classes of methods and derive properties of its regions of absolute stability. In section 4 the boundary of A_s is studied and a list of all possible boundary behaviours at the origin is given, see Table 1. In paragraph 5 we discuss the Daniel and Moore conjecture, see [6, p. 80]. This conjecture states, that the highest error order, p_{max}, a convergent A-stable (k,ℓ)-method can have is 2ℓ. Using techniques of Genin [8] we show that p_{max} has to be an even number. Moreover it is seen that the globally asymptotically exact methods play a key rôle in the determination of p_{max} . In the last section we draw some conclusions and give an outlook on work still to be done.

We shall use the sets

$$D(r) = \{\mu \in \mathbb{C} \mid \ |\mu| < r\} \ ,$$

$$H^+ = \{\mu \in \mathbb{C} \mid \ \mathrm{Re}\ \mu > 0\}$$

and

$$H^- = \{\mu \in \mathbb{C} \mid \ \mathrm{Re}\ \mu < 0\} \ .$$

2. Elementary properties of the stability regions and some further definitions.

First we consider the regions of relative stability. Clearly all three sets R_s, R_e and R_{es} are symmetric with respect to the real axis, R_s is open and one has $R_s \subset R_{es} \subset R_e$. The sets R_e and R_{es} differ only in finitely many points since an algebraic function has only finitely many branch points. Hence one has $\overline{R}_{es} = \overline{R}_e$. However in general $\overline{R}_s \neq \overline{R}_{es}$. Consider for example the convergent method given by $\Phi(\zeta,\mu) = \zeta^2 - 1 - 2\mu$. Here $\Omega = \mathbb{C} - (-\infty,-1/2]$ and one finds that $\zeta_1(\mu) = \sqrt{1+2\mu}$ and $\zeta_2(\mu) = -\sqrt{1+2\mu}$. Hence $R_s = \emptyset$, but $R_{es} = R_e = \Omega$. The roots $\zeta_i(0)$ of $\rho_0(\zeta)$ of modulus one are called essential roots. We shall always assume that the (k,ℓ)-method has s essential roots, $1 \leq s \leq k$.

Theorem 1: Assume the (k,ℓ)-method is convergent. Then one of the following four situations occurs:

(i) The method has no essential roots except $\zeta_1(\mu)$. Then the origin is contained in the interior of R_s .

(ii) The method has at least one essential root $\zeta_2(\mu) \neq \zeta_1(\mu)$. Then one of the following situations occurs:

(ii_a) $0 \in \partial R_s = \overline{R}_s - \mathring{R}_s$ and $0 \in \partial R_e = \overline{R}_e - \mathring{R}_e$. 1)

(ii_b) $R_s \cap D(\hat{r}) = \emptyset$ for $\hat{r}$ sufficiently small, $\hat{r} > 0$, and $0 \in \partial R_e$.

(ii_c) $R_s = \emptyset$ and $R_e = \Omega$.

Proof: Part (i) is obvious and shall be omitted. Part (ii). We have to compare $\zeta_1(\mu)$ with all branches $\zeta_i(\mu)$, $i = 2,3,\ldots,s$, which belong to essential roots $\zeta_i(0)$. Since the essential roots are simple and of modulus one there exists $r > 0$ such that the $\zeta_i(\mu)$ for $i = 1,2,\ldots,s$ are analytic and nonzero in $D(r)$. Hence

---

1) If S is a set in $\mathbb{C}$ then $\overline{S}$ denotes its closure and $\mathring{S}$ its interior.

$$h_i(\mu) = \frac{\zeta_1(\mu)}{\zeta_i(\mu)} \quad , \quad i = 2,3,\ldots,s$$

and $h_i^{-1}(\mu)$ are analytic in $D(r)$. Assume first that all $h_i(\mu)$ are constant in $D(r)$. Hence by analytic continuation $h_i(\mu)$ is constant in Ω and thus $R_s = \emptyset$ and $R_e = \Omega$. This is case (ii_c). Assume now that all $h_i(\mu)$ are not constant. To each $h_i(\mu)$ and $h_i^{-1}(\mu)$ we apply the maximum principle and find that to any $\tilde{r}$, $0 < \tilde{r} < r$ there exist μ's in $D(\tilde{r})$ with $|h_i(\mu)| > 1$ and some with $|h_i(\mu)| < 1$. Hence even if 0 is an isolated point of R_e one has always $0 \in \partial R_e$. Since the $h_i(\mu)$ are locally conformal maps at the origin the curves given by $|h_i(\mu)| = 1$ have a tangent at the origin. Hence either $0 \in \partial R_s$ or $R_s \cap D(\hat{r}) = \emptyset$ for $\hat{r}$ sufficiently small, $\hat{r} > 0$. This covers the case (ii_a) completely and (ii_b) partly. If some of the $h_i(\mu)$ are constant and others are not, one finds the remaining possibilities which lead to case (ii_b). □

Note that since the essential roots $\zeta_i(0)$ are simple there exists a disk $D(r)$, $r > 0$, r sufficiently small such that R_e and R_{es} restricted to $D(r)$ are equal.

We turn now our attention to the regions of absolute stability. Since the coefficients of $\Phi(\zeta,\mu)$ are real the sets A_s, A_e and A_{es} are symmetric with respect to the real axis. From the definitions it is evident that $A_s \subset A_{es} \subset A_e$ and A_s is an open set. The sets A_{es} and A_e differ only in finitely many points. Hence $\overline{A}_{es} = A_e = \overline{A}_e$. However in general $\overline{A}_s \neq A_e$. For example for the Milne-Simpson method one finds $A_s = \emptyset$, $A_{es} = \{iy \mid |y| < 3\}$ and $A_e = \overline{A}_{es}$. In the following we restrict ourselves to deal with A_s only. Similar results with obvious modifications usually hold for A_e and A_{es} too. To determine A_s one needs the set

(11) $\Gamma = \{\mu \in \mathbb{C} \mid$ There exists at least one branch $\zeta_i(\mu)$ with $|\zeta_i(\mu)| = 1\}$.

It can be shown, see Jeltsch [14, p. 4.2-6 ff] or Rubin [17, p. 34] , that Γ consists of a finite number of analytic curves which intersect each other in finitely many points. The complement of Γ consists of finitely many connected components $\Omega_1,\Omega_2,\ldots,\Omega_t$. Let $\mu_i \in \Omega_i$ then $\Omega_i \subset A_s$ if and only if $\mu_i \in A_s$ and if $\mu_i \notin A_s$ then $\Omega_i \cap A_s = \emptyset$. This provides a way of computing A_s . For $\zeta = e^{i\phi}$, $\phi \in [0,\pi]$ one computes the roots of $\Phi(e^{i\phi},\mu)$. This gives a set C and then $\Gamma = C \cup C^c$, where $C^c = \{\mu \mid \overline{\mu} \in C\}$. Once Γ is known one tests for each component Ω_i of the complement of A_s whether it belongs to A_s or not by testing just one sample point $\mu_i \in \Omega_i$. We shall need the following definitions.

<u>Definition 1</u>: (Griepentrog [9]) A (k,ℓ)-method is called <u>globally asymptotically exact</u> if $A_s = H^-$.

Methods of this form reflect the asymptotic behaviour of the exact solution of (3) correctly as n tends to ∞ and h remains fix. The trapezoidal rule is a well known example. Usually the requirement of global asymptotic exactness is to strong and one weakens it to request only that $H^- \subset A_s$. Methods with this property are called <u>A-stable</u>, see Dahlquist [5].

<u>Definition 2</u>: (Griepentrog [9]) A (k,ℓ)-method is called <u>asymptotically exact</u> if there exists a disk $D(r)$, $r > 0$, r sufficiently small, such that

$$A_s \cap D(r) = H^- \cap D(r) \quad .$$

Even though globally asymptotically exact methods and asymptotically exact methods may not be needed in practicle applications to solve stiff differential equations they may be the key link in the proof of the Daniel and Moore conjecture, see section 5. It can be shown, see Jeltsch [14, p. 4.4-1] that a convergent asymptotically exact $(k,1)$-method is globally asymptotically exact. However this is not true for a general (k,ℓ)-method with $\ell > 1$. A counterexample is the $(1,2)$-method whose region of absolute stability is given in Fig. 2a.

For easier reference we need the polynomials

$$(12) \qquad \rho_j(\zeta) = \sum_{i=0}^{k} \alpha_{ij}\, \zeta^i \;, \quad j = 0,1,\ldots,\ell \quad ;$$

and

$$(13) \qquad \eta_i(\mu) = \sum_{j=0}^{\ell} \alpha_{ij}\, \mu^j \;, \quad i = 0,1,\ldots,k \quad .$$

<u>Definition 3</u>: A (k,ℓ)-method is called symmetric if

$$(14) \qquad \eta_i(\mu) = -\eta_{k-i}(-\mu), \; i = 0,1,\ldots,k \quad .$$

To each essential root $\zeta_i(0)$ one introduces the growth parameters

$$\lambda_i = - \frac{\rho_1(\zeta_i(0))}{\zeta_i(0)\rho_0'(\zeta_i(0))}$$

as it is done for $(k,1)$-methods, see e.g. Henrici [10, p. 237]. One can show that a convergent symmetric (k,ℓ)-method has an even error order p, k essential roots and all growth parameters are real. The proof runs along the lines of the proof of the special case $\ell = 1$ which is treated in Stetter [19, p. 251].

3. Symmetric methods and the symmetry principle.

Theorem 2. Assume a (k,ℓ)-method is convergent and $\Phi(\zeta,\mu)$ is irreducible. Then the following two statements are equivalent:

(A) For some r , $r > 0$ one has $I = \{iy \mid |y| < r\} \subset \Gamma$.

(B) The method is symmetric.

Proof (A) $\Rightarrow$ (B). Let $\zeta_1(0),\ldots,\zeta_s(0)$, be the essential roots. Since these are simple the corresponding branches of $\zeta(\mu)$ given by (7) are analytic and even conform in a disk $D(r)$, $r > 0$, r sufficiently small. Further assume that r is chosen so small that for the nonessential branches $\zeta_i(\mu)$, $i > s$, one has $|\zeta_i(\mu)| < 1$ for all $\mu \in D(r)$. Since $I \subset \Gamma$ there exists at least one branch $\zeta_t(\mu)$ and r_t , $0 < r_t < r$ such that $|\zeta_t(iy)| = 1$ for all $|y| < r_t$. $\zeta_t(\mu)$ maps $I_t = \{iy \mid |y| < r_t\}$ on to a segment C_t of the circle $|\zeta| = 1$. Since the map by $\zeta_t(\mu)$ is conform we can apply the symmetry principle of Schwarz (see e.g. Ahlfors [1,171]) and find

$$\zeta_t(-\bar{\mu}) = \overline{\zeta_t(\mu)}^{\,-1}. \tag{15}$$

From $\sum_{i=0}^{k} \eta_i(\mu)\zeta^i = 0$ follows

$$\sum_{i=0}^{k} \eta_i(-\bar{\mu})\overline{\zeta_t(\mu)}^{\,-i} = 0 \ . \tag{16}$$

Using analytic continuation and the reflection principle by Schwarz one can show that for any $\mu \in D(r_t)$ and ζ with

$$\sum_{i=0}^{k} \eta_i(\mu)\zeta^i = 0 \tag{17}$$

one has

$$\sum_{i=0}^{k} \eta_i(-\bar{\mu})\bar{\zeta}^{-i} = 0 \ .^{1)} \tag{18}$$

Take the conjugate complex of (18) and multiply it by ζ^k to obtain

$$\sum_{i=0}^{k} \overline{\eta_{k-i}(-\bar{\mu})}\zeta^i = \zeta^k \overline{\Phi(\bar{\zeta}^{-1},-\bar{\mu})} = 0 \ . \tag{19}$$

Clearly $\zeta^k \overline{\Phi(\bar{\zeta}^{-1},-\bar{\mu})}$ is irreducible. For each $\mu \in D(r_t)$ the polynomials (17) and (19) have exactly the same roots. Hence there exists a constant c , $c \neq 0$, such that

$$\eta_i(\mu) = c\,\overline{\eta_{k-i}(-\bar{\mu})} = c\,\eta_{k-i}(-\mu) \ . \tag{20}$$

Since the method is convergent all roots of $\rho_0(\zeta)$ have to be of modulus one for

---

1) It should be noted, that if one does not assume that the method is convergent then ζ and $\bar{\zeta}^{-1}$ may in general belong to different branches. In particular to $\mu = 0$ may belong a ζ with $|\zeta| < 1$ and hence $|\bar{\zeta}^{-1}| > 1$.

otherwise not only the pair $\mu = 0$, ζ_i with $|\zeta_i| < 1$ would satisfy (17) but also the pair $\mu = 0$, $\zeta = \bar{\zeta}_i^{-1}$. This is however a contradiction to the convergence of the method. By Vieta's theorem we find $\alpha_{k0}/\alpha_{00} = -1$. Hence $c = -1$ and the symmetry of the method is established.

(B) ⇒ (A). From the symmetry and stability follows that the method has only essential roots. Hence, at $\mu = 0$ we have k different branches $\zeta_i(\mu)$ which are analytic in $D(r)$, $r > 0$, r suffieciently small. By assumption

$$\eta_i(\mu) = - \eta_{k-i}(-\mu) \ , \ i = 0,1,\dots,k \ .$$

and hence

$$\Phi(\zeta,\mu) = -\zeta^k \, \Phi(\zeta^{-1},-\mu) \ . \tag{21}$$

Let $\zeta_t(\mu)$ be a branch of $\zeta(\mu)$. Then by (21)

$$\Phi(\bar{\zeta}_t^{-1}(\mu),-\bar{\mu}) = 0$$

for all $\mu \in D(r)$. Therefore $\zeta = \bar{\zeta}_t^{-1}(-\bar{\mu})$ is a branch of $\zeta(\mu)$, too. We distinguish two cases.

(i) There exists a branch $\zeta_j(\mu)$, $j \neq t$ with

$$\zeta_j(\mu) = \bar{\zeta}_t^{-1}(-\bar{\mu}) \quad \text{for all} \quad \mu \in D(r) \ .$$

In particular

$$\zeta_j(0) = \bar{\zeta}_t^{-1}(0) = \zeta_t(0) \ .$$

This is a contradiction to $\zeta_i(0) \neq \zeta_j(0)$ for $i \neq j$.

(ii) $\zeta_t(\mu) = \bar{\zeta}_t^{-1}(-\bar{\mu})$ for all $\mu \in D(r)$. In particular, let $\mu = iy$, $|y| < r$. Then

$$\zeta_t(iy) = \bar{\zeta}_t^{-1}(iy)$$

and

$$|\zeta_t(iy)| = 1 \quad \text{for} \quad |y| < r \ , \quad \text{and} \quad t = 1,2,\dots,k \ . \ \square$$

It should be remarked that we have proved in the second part not only that $I \subset \Gamma$ but $I \subset A_{es}$. We shall further need that an essential branch has at the origin the following representation (see e.g. Jeltsch [12])

$$\zeta_i(\mu) = \zeta_i(0)(1 + \lambda_i \, \mu \ + O(\mu^2)) \ .$$

These two facts lead to the following corollaries.

<u>Corollary 1</u>. Assume a (k,ℓ)-method is convergent and $\Phi(\zeta,\mu)$ is irreducible. Then the following two statements are equivalent:

(C) $A_s \cap D(r) = \emptyset$, $A_{es} \cap D(r) = \{iy \mid |y| < r\}$ for some $r > 0$, r sufficiently small.

(D) The method is symmetric and at least one growth parameter λ_i is negative.

<u>Corollary 2</u>. Assume a (k,ℓ)-method is convergent and $\Phi(\zeta,\mu)$ is irreducible. Then the following two statements are equivalent:

(E) The method is asymptotically exact.

(F) The method is symmetric and all growth parameters are positive.

To conclude this section we give the symmetry principle of Γ. First notethat μ and $\bar{\mu}$ are symmetric with respect to the real axis.

Definition 4. The linear transformation

$$(22) \qquad T : z \to w = \frac{az + b}{cz + d} \quad , \quad a,b,c,d \in \mathbb{C}$$

carries the real axis into a circle C or a straight line L. Then $w = T\mu$ and $w^* = T\bar{\mu}$ are said to be symmetric with respect to the circle C, the straight line L respectively.

The such defined relation between w, w^* and C, L respectively, is independent of T, see Ahlfors [1,p. 27]. A set S is said to be symmetric with respect to a circle C, a straight line L respectively if to each $\mu \in S$ its symmetric point μ^* is in S too.

Theorem 3: (Symmetry principle of Γ) Let $\Phi(\zeta,\mu)$ be irreducible and Γ be given by (11). Assume there exists a segment γ of a straight line L or a circle C with $\gamma \subset \Gamma$. Then Γ is symmetric with respect to L or C respectively.

Outline of the proof. There is a point μ_0 on γ where $\zeta(\mu)$ given by (7) has no branch points. There exists a linear transformation $T : \mu \to w = T\mu$ of form (22) which maps γ onto the imaginary axis with $T\mu_0 = 0$ and is one to one on $\mathbb{C}$, see Ahlfors [1,p. 23]. Let $\tilde{\zeta}(w)$ be the algebraic function defined by

$$\chi(\tilde{\zeta}(w),w) = \Phi(\zeta(w),T^{[-1]}w) = 0 \quad ,$$

where $T^{[-1]}$ is the inverse map of T. Let Γ_w and γ_w be the images of Γ and γ respectively. Then

$$\Gamma_w = \left\{ w \in \mathbb{C} \;\middle|\; \begin{array}{l} \text{branch } \tilde{\zeta}_i(w) \text{ of } \tilde{\zeta}(w) \\ \text{with } |\zeta_i(w)| = 1 \end{array} \right\} \quad .$$

The algebraic function $\tilde{\zeta}(w)$ satisfies now the assumption (A) in Theorem 2 except that at $w = 0$ one has in general $|\tilde{\zeta}_i(0)| \neq 1$ for some i. Nevertheless we can, as in the first part of the proof of Theorem 2, show that

$$(23) \qquad \chi(\zeta,w) = c\zeta^k \, \overline{\chi(\bar{\zeta}^{-1},-\bar{w})}$$

Let $\hat{w} \in \Gamma_w$. Hence there exists $\hat{\zeta}$ with $|\hat{\zeta}| = 1$ and $\chi(\hat{\zeta},\hat{w}) = 0$. From (23) it follows that

$$\chi(\bar{\hat{\zeta}}^{-1},-\bar{\hat{w}}^{-1}) = 0 \quad .$$

Thus $-\bar{\hat{w}} \in \Gamma_w$ since $|\bar{\hat{\zeta}}^{-1}| = 1$. Hence Γ_w is symmetric with respect to the imaginary axis. Γ is the image of Γ_w induced by the map $T^{[-1]}$. However $T^{[-1]}$ maps points symmetric to the imaginary axis into points symmetric to C or L, respectively. Hence Γ is symmetric with respect to C or L, respectively. □

As an illustration we show in Fig. 2 two Γ which contain a segment of a circle.

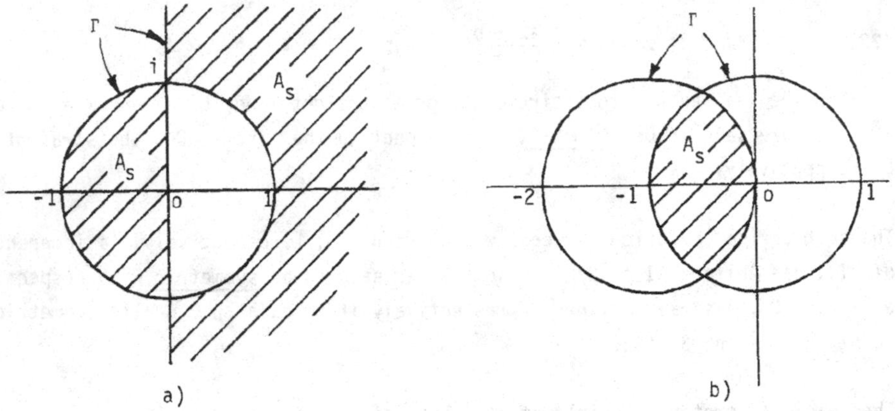

Fig. 2.

Fig. 2a depicts Γ and A_s of the method $y_{n+1} - y_n = \frac{h}{2}(f_{n+1}+f_n) - h^2(-f_{n+1}^{(1)}+f_n^{(1)})$ while in Fig. 1b the method is $y_{n+2} - y_{n+1} = h(2f_{n+1}-f_n) - h^2 f_n^{(1)}$. Clearly Γ in Fig. 2b is not symmetric with respect to $\partial D(1)$ even so Γ contains a segment of a circle. This is not a contradiction to Theorem 2 since the assumption that $\Phi(\zeta,\mu)$ is irreducible is violated. In fact

$$\begin{aligned}\Phi(\zeta,\mu) &= \zeta^2 - \zeta(1 + 2\mu) + \mu^2 + \mu \\ &= (\zeta - (1 + \mu))(\zeta - \mu) .\end{aligned}$$

4. The boundary behaviour of the region of absolute stability.

Recall that the complement of Γ consists of finitely many connected components $\Omega_1, \Omega_2, \ldots, \Omega_t$ and that one either has $\Omega_i \subset A_s$ or $\Omega_i \cap A_s = \emptyset$. Hence ∂A_s is a subset of Γ. ∂A_s is a piecewise smooth curve with finitely many edges or cusps. Each edge μ^* has a uniquely defined positive angle α^*. This angle can be characterized using the branches $\zeta_i(\mu)$ at μ^* or the coefficients of $\Phi(\zeta,\mu)$. For details, see Jeltsch[13] , [14] . In particular one finds that α^* can never

Table 1

Fig. of A_{es}	Comments
a) 	$A_{es} \cap D = \{0\}$, $A_s \cap D = \emptyset$. SC : Re $\lambda_i < 0$ for some i
b) 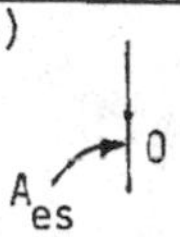	$A_{es} \cap D = \{iy \mid \lvert y\rvert < r\}$, $A_s \cap D = \emptyset$. Characterized by Corollary 1. Ex: Milne-Simpson method
c) 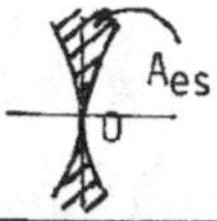	NC: All λ_i real and at least one negative Ex: $2y_{n+3} - y_{n+2} - 2y_{n+1} + y_n = h(-f_{n+3} + 6f_{n+2} - f_{n+1} - 2f_n)$
d) 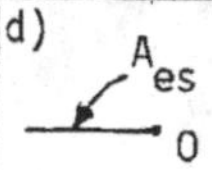	NC: $\mathrm{Min}(\pi/2 - \arg \lvert\lambda_i\rvert) = 0$. Ex: Formula (24)
e) 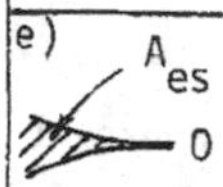	NC: $\mathrm{Min}(\pi/2 - \arg \lvert\lambda_i\rvert) = 0$. Ex: $y_{n+3} - y_{n+2} + y_{n+1} - y_n = h(2f_{n+3} + f_{n+1} - f_n)$
f) 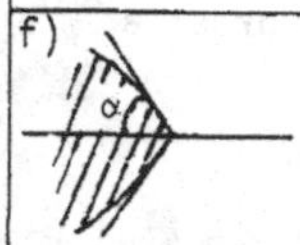	NC and SC: $\alpha = \min(\pi/2 - \arg \lvert\lambda_i\rvert)$ This generalizes Theorem 4.6.4 in Stetter [19, p. 267]
g) 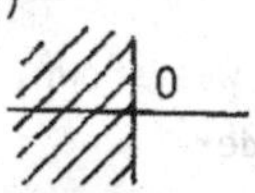	Characterized by Corollary 2.
h) 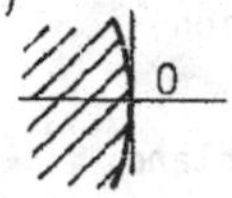	NC: $\lambda_i > 0$ for all i . For further NC and SC see Jeltsch [12] .
i) 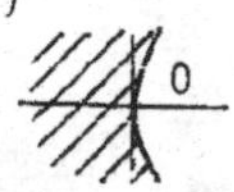	NC: $\lambda_i > 0$ for all i . For further NC and SC see Jeltsch [12] .

Abreviations used in Table 1: Ex = Example, NC = Necessary Condition,
SC = Sufficient condition, D = disk $D(r)$,
r sufficiently small , $r > 0$.

exceed π . It may happen that two edges come together. This happens for example at $\mu^* = i$ in Fig. 2a. For convergent methods one has some further information about the branches at the origin. Hence we shall classify the behaviour of ∂A_s at the origin. In each case we consider only ∂A_s in a disk $D = D(r)$ where r is sufficiently small, $r > 0$. The different possibilities are given in Table 1. In all cases convergent (k,ℓ)-methods with $\Phi(\zeta,\mu)$ irreducible and the shape of A_{es} as indicated are known except for case d). Here one has the example

$$(24) \qquad y_{n+3} - y_{n+2} + y_{n+1} - y_n = h(f_{n+3} - 4f_{n+2} + 5f_{n+1}) + h^2(-f^{(1)}_{n+3} + 5f^{(1)}_{n+2} - f^{(1)}_{n+1} + f^{(1)}_n) + h^3(f^{(2)}_{n+3} + f^{(2)}_{n+1}) \ .$$

$\Phi(\zeta,\mu)$ is reducible since

$$\Phi(\zeta,\mu) = \{\zeta(1-\mu)-1\}\{\zeta^2(1+\mu^2) + 4\mu\zeta + 1 + \mu^2\} \ .$$

5. The Daniel and Moore conjecture.

So far we have just considered the region of absolute stability A_s of a convergent (k,ℓ)-method. In this section we relate the error order p of a method to one particular shape of A_s , namely $H^- \subset A_s$. Let p_{max} be the highest error order a convergent, A-stable, (k,ℓ)-method can have. Then one has the following

Conjecture (Daniel and Moore [6])

$$p_{max} = 2\ell \ .$$

It is well-known, see e.g. Ehle [7], that $p_{max} \geq 2\ell$. In Genin [8] it was shown that there exist unstable methods with $H^- \subset A_s$ which have the error order $p = 2\ell + \min\{k,\ell\} - 1$. We shall use the techniques of Genin [8] to show that p_{max} is an even number. It is convenient to use the variable transformation

$$(25) \qquad z = (\zeta+1)/(\zeta-1) \ ; \ \zeta = (z+1)/(z-1) \ .$$

This transformation maps the unit disk of the ζ-plane into the left hand plane. Moreover let $w = -\mu$. The polynomial $\rho_j(\zeta)$ are now transformed into

$$(26) \qquad R_j(z) = (-1)^j(z-1)^k\rho_j\left(\frac{z+1}{z-1}\right) = \sum_{i=0}^{k} A_{ij}z^i \ , \ j = 0,1,\dots,\ell \ ,$$

and (7) becomes the form

$$(27) \qquad H(z,w) = (z-1)^k\Phi\left(\frac{z+1}{z-1},-w\right) = \sum_{i=0}^{k}\sum_{j=0}^{\ell} A_{ij}z^iw^j \ .$$

If the method is convergent then $\rho_0(1) = 0$, $\rho_0'(1) \neq 0$, $\rho_1(1) \neq 0$. Since the transformation (25) maps $\zeta = 1$ into $z = \infty$ we have for a convergent method that

$$(28) \qquad A_{k0} = 0 \ , \ A_{k-1\,0} \neq 0 \ , \ A_{k1} \neq 0 \ .$$

Moreover for a stable method $R_0(z)$ has all roots in $\overline{H^-}$ and the purely imaginary roots are simple.

Definition 5 (Ansell [2, p. 219]) A polynomial in two variables $H(z,w)$ with real coefficients is called a two variable Hurwitzpolynomial in the narrow sense if it has zeros in neither $\operatorname{Re} z > 0$, $\operatorname{Re} w > 0$ nor $\operatorname{Re} z > 0$, $\operatorname{Re} w = 0$ nor $\operatorname{Re} z = 0$, $\operatorname{Re} w > 0$.

Lemma 1 (Genin [8]) A convergent (k,ℓ)-method is A-stable if and only if $H(z,w)$ is a two variable Hurwitz polynomial in the narrow sense.

We show in the following that among the convergent, A-stable (k,ℓ)-methods with $p = p_{max}$, there is at least one symmetric method. Let us define

(29a) $$H_e(z,w) = 1/2\,[\,H(z,w) + H(-z,-w)\,]$$

and

(29b) $$H_o(z,w) = 1/2\,[\,H(z,w) - H(-z,-w)\,]$$

the even and odd parts of $H(z,w)$.

Lemma 2 (Genin [8]) If a convergent (k,ℓ)-method is A-stable, $H(z,w)$ is irreducible and $H_o(z,w) \neq 0$ and $H_e(z,w) \neq 0$ then

(30a) $$H_o(z,w) = H_{o,r}(z,w) \prod_{i=1}^{u} (w-w_j) \prod_{i=1}^{v} (z-z_i)$$

and

(30b) $$H_e(z,w) = H_{e,r}(z,w) \prod_{j=1}^{u'} (w-w_j') \prod_{i=1}^{v'} (z-z_i') ,$$

where $\operatorname{Re} w_j = \operatorname{Re} w_j' = \operatorname{Re} z_i = \operatorname{Re} z_i' = 0$ and $H_{o,r}(z,w)$ and $H_{e,r}(z,w)$ are two variable Hurwitz polynomials in the narrow sense.

Since $H_o(z,w)$ has real coefficients, the w_j and z_i will occur in conjugate complex pairs or $w_j = 0$, $z_i = 0$. Hence $H_{o,r}(z,w)$ will either be odd or even. The same is true for $H_{e,r}(z,w)$. We need the following

Lemma 3 (Genin [8, p. 389]) A (k,ℓ)-method has error order p if and only if

(31) $$z^{-k} H(z,-\log\tfrac{z+1}{z-1}) = C_{p+1}(\tfrac{2}{z})^{p+1} + O(z^{-p-2}) \quad \text{as} \quad z \to \infty$$

where $C_{p+1} \neq 0$.

Note that $\log((z+1)/(z-1))$ is an odd function with a simple zero at $z = \infty$. We are now in a position to prove the following

Theorem 4. To find p_{max} for a given pair of natural numbers k and ℓ it is enough to consider all convergent, A-stable (k',ℓ')-methods with $H(z,w)$ irreducible and $H(z,w)$ even if k' is odd and $H(z,w)$ odd if k' is even for $1 \le k' \le k$ and $1 \le \ell' \le \ell$.

Proof: Consider a convergent, A-stable (k,ℓ)-method with $H(z,w)$ irreducible. We distinguish the following two cases

(I) k even. From (28) follows that $H_o(z,w) \not\equiv 0$. The expansion (31) can be split in the two parts

(32a) $$z^{-k} H_e(z,-\log(\tfrac{z+1}{z-1})) = \sum_{s=t}^{\infty} \gamma_{2s+2} z^{-2s}$$

and

(32b) $$z^{-k} H_o(z,-\log(\tfrac{z+1}{z-1})) = \sum_{s=u}^{\infty} \gamma_{2s+1} z^{-2s-1} .$$

We consider the new method based on

(33) $$\tilde{H}(z,w) = H_o(z,w) .$$

Note that $\tilde{H}(z,w)$ still satisfies (28), a fact which is not true if one would have used $\tilde{H}(z,w) = H_e(z,w)$. From (32b) follows that the error order $\tilde{p}$ of the new method $\tilde{H}(z,w)$ is even; in fact

(34) $$\tilde{p} = \begin{cases} p & \text{if } p \text{ is even,} \\ p + 1 + 2t & \text{is } p \text{ is odd ,} \end{cases}$$

where p is the error order of the original method and t is some nonnegative integer. $\tilde{H}(z,w)$ might be reducible. Consider therefore the method given by

$$\hat{H}(z,w) = H_{o,r}(z,w) ,$$

where $H_{o,r}(z,w)$ is given in (30a). From Lemma 2 and Lemma 1 follows that the method given by $\hat{H}(z,w)$ is A-stable. Since $\hat{H}(z,w)$ still satisfies (28), the error order of the corresponding method is $\hat{p} = \tilde{p}$. Since $H_o(z,w)$ was odd $\hat{H}(z,w)$ will either be even or odd. $\hat{H}(z,w)$ still might be truly reducible. Exactly one of the factors, call it $H'(z,w)$ will satisfy (28). Since $\hat{H}(z,w)$ was a two variable Hurwitz polynomial in the narrow sense $H'(z,w)$ has to be one too. Hence the corresponding method is A-stable and stable. Moreover the error order $p' = \hat{p} = \tilde{p}$ and $H'(z,w)$ is either even or odd. From (28) follows that k' is even if $H'(z,w)$ is odd and k' is odd if $H'(z,w)$ is even. Obviously $k' \le k$.

(II) k is odd. From (28) follows that $H_e(z,w) \not\equiv 0$. One proceeds now in an analog manner as in case (I) by setting

$$\tilde{H}(z,w) = H_e(z,w)$$

and finds

(35) $$\tilde{p} = \begin{cases} p & \text{if } p \text{ is even} \\ p + 1 + 2t' & \text{if } p \text{ is odd .} \end{cases}$$

The rest of the details are left to the reader. □

Corollary 3. p_{max} is even.
This follows directly from (34) and (35).

Corollary 4. To find p_{max} to a given pair of natural numbers k and ℓ it is enough to consider all convergent globally asymptotically exact (k',ℓ')-methods with $1 \le k' \le k$, $1 \le \ell' \le \ell$ and $H(z,w)$ irreducible.

Note that Theorem 4 is stronger than the corresponding theorem given by Genin [8]. This is due to the fact that Genin admits nonconvergent methods, that is, in the reduction process in [8] it is not ensured that (28) remains satisfied. Corollary 4 shows the key rôle which is played by the globally asymptotically exact methods in the problem of determining p_{max}.

Theorem 5. The Daniel and Moore conjecture is correct for $k + \ell \le 5$ and $k = 2$, $\ell = 4$.

Proof: In this proof one needs results of Dahlquist [4], Genin [8], Jeltsch [11] and Reimer [16]. Since for each pair of k and ℓ the arguments are almost the same we just treat one typical case, namely $k = 2$, $\ell = 2$. In [4], [11], [16] it is shown that the highest error order of a convergent (2,2)-method is 6. Genin [8] computed $H(z,w)$ of the (2,2)-methods of order 6 and found

$$H(z,w) = [30z + (15z^2-1)w + 2zw^2]\alpha + [48 + 18z,w - (3z^2 - 5)w^2]\beta \tag{36}$$

where α and β are free real parameters. However Genin [8] showed that a necessary condition for A-stability is that $A_{ij} > 0$ for all i and j. For any choice of α and β this will be violated in (36). Hence $p_{max} \le 5$. But by Corollary 3 we know that p_{max} has to be even. Hence $p_{max} \le 4$. However from $p_{max} \ge 2\ell = 4$, see e.g. Ehle [7], we find $p_{max} = 4$. The other cases are treated similarly, for details, see Jeltsch [14]. □

6. Conclusions

Even though we have restricted ourselves to (k,ℓ)-methods many results can be applied to other methods too, such as composite multistep methods, see Rubin [17], cyclic multistep methods, see Stetter [19], Runge-Kutta methods and the large class of methods introduced by Nevanlinna, Sipilä [15]. In all these methods the stability of the method when applied to $y' = \lambda y$ can be characterized by a two variable polynomial $\Phi(\zeta,\mu)$ in the same manner as with (k,ℓ)-methods. However in the other methods the coefficients of $\Phi(\zeta,\mu)$ have to satisfy some further conditions which are due to the special form of the method. For example there exists a corresponding conjecture to the Daniel and Moore conjecture for composite multistep methods namely $p_{max} \le \min\{2\text{rank } B_f, 2\ell\}$. Here the bound $2 \text{ rank } B_f$ is due to the side condition the coefficients of $\Phi(\zeta,\mu)$ have to satisfy and the bound 2ℓ is basically the Daniel and Moore conjecture. For details see Bickert, Burgess and Sloate [3] and Sloate and Bickert [18].

A general line to follow in the research involving regions of absolute stability would be the following. Let M denote a set of (k,ℓ)-methods of which A_s satisfies a certain property, e.g. A_0-stable methods, $A(\alpha)$-stable methods, stiffly stable methods, A-stable methods e.t.c.. For each of such a class of methods one is interested in the following problems:

(i) Characterize all (k,ℓ)-methods in M.

If this is too difficult to solve then one tries

(ii) Find necessary conditions for a (k,ℓ)-method to be in M.

and

(iii) Find sufficient conditions for a (k,ℓ)-method to be in M.

Furthermore one would like to answer the questions:

(iv) For a given k,ℓ and M what is the highest order of a (k,ℓ)-method in M?

or

(v) For a given s,ℓ and M what is the highest "s-value" multistep method using ℓ derivatives in M, when $\ell > s$? s stands here for the number of backinformation which has to be stored. For example in the Adams-Bashforth method $s = k + 1$ while in the optimal $(k,1)$-methods of Dahlquist, see Henrici [10, p. 233] one has $s = 2k$.

The same questions can be asked when the class M is taken to be a subclass of some type of methods such as composite multistep methodsor cyclic methods.

Finally we would like to caution the reader that the region of absolute stability may sometimes look as if the methods are good for solving stiff differential equations but in fact they are not. As an example consider the (3,1)-method of order 3 with $\rho_1(\zeta) = (15 - 17\zeta)^3$. For this method one has $\{\mu \in \mathbb{C} \mid \mathrm{Re}\,\mu < -0.0012\} \subset A_s$. But while solving a stiff system with one eigenvalue of the Jacobian approximately -4000 from $x = 0$ to $x = 50$ with a fixed step $h = 1$ we observed a relative error of 4. There are various reasons for this behaviour but an analysis of these phenomenas would go beyond the scope of this article.

References

1. Ahlfors, L.V., Complex Analysis, McGraw-Hill, New York, 1953.

2. Ansell, H.G., On certain two-variable generalizations of circuit theory, with applications to networks of transmission lines and lumped reactances, IEEE Trans. on C.T. 11, (1964), 214-223.

3. Bickart, T.A., D.A. Burgess and H.M. Sloate, High order A-stable composite multistep methods for numerical integration of stiff differential equations, in Proc. 9th Annual Allerton Conf. on Circuit and System Theory, (1971), 465-473.

4. Dahlquist, G., Convergence and stability in the numerical integration of ordinary differential equations, Trans. Roy. Inst. Tech., Stockholm, Nr. 130, 1959.

5. -------------, A special stability problem for linear multistep methods, BIT 3, (1963), 27-43.

6. Daniel, J.W. and R.E. Moore, Computation and theory in ordinary differential equations, Freeman and Co., San Francisco, 1970.

7. Ehle, B.L., High order A-stable methods for the numerical solution of systems of D.E.'s, BIT 8, (1968), 276-278.

8. Genin, Y., An algebraic approach to A-stable linear multistep-multiderivative integration formulas, BIT 14, (1974), 382-406.

9. Griepentrog, E., Mehrschrittverfahren zur numerischen Integration von gewöhnlichen Differentialgleichungssystemen und asymptotische Exaktheit, Wiss. Z. Humboldt-Univ. Berlin Math.-Natur. Reihe, v. 19, (1970), 637-653.

10. Henrici, P., Discrete variable methods in ordinary differential equations, Wiley, New York, 1962.

11. Jeltsch, R., Integration of iterated integrals by multistep methods, Numer. Math. 21, (1973), 303-316.

12. Jeltsch, R., A necessary condition for A-stability of multistep multiderivative methods, to appear in Math. Comp., 30 (1976).

13. Jeltsch, R., Stiff stability of multistep multiderivative methods, to appear in SIAM J. on numer. Anal.

14. Jeltsch, R., Multistep multiderivative methods for the numerical solution of initial value problems of ordinary differential equations., Seminar Notes 1975/76, University of Kentucky, 1976.

15. Nevanlinna, O. and A.H. Sipilä, A nonexistence theorem for explicit A-stable methods, Math. Comp., 28 (1974), 1053-1055.

16. Reimer, M., Finite difference forms containing derivatives of higher order, SIAM J. Numer. Anal., 5 (1968), 725-738.

17. Rubin, W.B., A-stability and composite multistep methods, Ph. D. Thesis, EE Dept., Syracuse University, New York, 1973.

18. Sloate, H.M. and T.A. Bickart, A-stable composite multistep methods, JACM 20, (1973), 7-26.

19. Stetter, H.J., Analysis of discretization methods of ordinary differential equations, Springer, New York, 1973.

Rolf Jeltsch
Institute of Mathematics
Ruhr-University Bochum
D-4630 Bochum
Federal Republic of Germany

PRÄDIKTOREN MIT VORGESCHRIEBENEM STABILITÄTSVERHALTEN

R. Mannshardt

Rechenzentrum
der Ruhr-Universität Bochum
Universitätsstraße 150

D-4630 Bochum

1. Einleitung

Integriert man ein System von gewöhnlichen Differentialgleichungen mit einem impliziten linearen Mehrschrittverfahren, so muß man in jedem Integrationsschritt ein (lineares oder nichtlineares) Gleichungssystem auflösen. Bei den üblichen Prädiktor-Korrektor-Verfahren geschieht das durch einen Iterationsprozeß, der nur konvergiert, wenn die Schrittweite hinreichend klein ist. Diese Einschränkung der Schrittweite kann bewirken, daß die besonderen Stabilitätseigenschaften (z. B. A-Stabilität) des impliziten Verfahrens praktisch nicht ausgenützt werden.

Im folgenden wird zunächst gezeigt, wie man zu einem impliziten Verfahren (als Korrektor) einen Prädiktor konstruiert, der ein vorgeschriebenes Stabilitätsverhalten besitzt. Das zugehörige $P(EC)^mE$-Verfahren übernimmt dieses Stabilitätsverhalten und vermeidet die oben genannte Einschränkung der Schrittweite. Eine Analyse der Stabilitätsbereiche und des lokalen Verfahrensfehlers ergibt, daß speziell das PECE-Verfahren für gewisse steife Systeme geeignet ist. Diese Untersuchungen werden dann auf Verfahren mit höheren Ableitungen übertragen. Zum Schluß werden die theoretischen Überlegungen anhand eines Beispiels veranschaulicht, bei dem das PECE-Verfahren wesentlich weniger Rechenzeit benötigt als der durch Newton-Iteration aufgelöste Korrektor.

2. Stabilitätspolynome und Nullbedingung

Ein Korrektor sei bestimmt durch die charakteristischen Polynome

$$(2.1)\qquad \rho(z) = \sum_{i=0}^{k} \alpha_i z^i \text{ mit } \alpha_k=1, \quad \sigma(z) = \sum_{i=0}^{k} \beta_i z^i \quad (\beta_k \neq 0)$$

und ein Prädiktor durch

$$(2.2)\qquad \rho^*(z) = \sum_{i=0}^{k} \alpha_i^* z^i \text{ mit } \alpha_k^*=1, \quad \sigma^*(z) = \sum_{i=0}^{k-1} \beta_i^* z^i .$$

(Ist der Korrektor oder der Prädiktor ursprünglich ein j-Schritt-Verfahren mit $j<k$, so multipliziere man seine charakteristischen Polynome mit

z^{k-j}, um obige Darstellung zu erhalten.) Bei Anwendung auf die Testgleichung $y'=\lambda y(\lambda\in\mathbb{C})$ mit der Schrittweite h erhält man die zugehörigen Stabilitätspolynome mit $\bar{h}:=h\lambda$:

$$(2.3) \quad \pi(z,\bar{h}) = \rho(z)-\bar{h}\sigma(z), \qquad \pi^*(z,\bar{h}) = \rho^*(z)-\bar{h}\sigma^*(z) .$$

Das Stabilitätspolynom des $P(EC)^mE$-Verfahrens (vgl. [6], S. 97, [12] S. 326 ff) kann mit $\bar{q}:=\bar{h}\beta_k$ in folgender Gestalt geschrieben werden (falls $\bar{q}\neq 1$):

$$(2.4) \quad \overset{m}{\pi}(z,\bar{h}) = \frac{1}{1-\bar{q}} \left(\pi(z,\bar{h}) + \hat{\pi}(z,\bar{h})\,\bar{q}^m\right)$$

mit der Hilfsfunktion

$$(2.5) \quad \hat{\pi}(z,\bar{h}) = -\pi(z,\bar{h}) + (1-\bar{q})\pi^*(z,\bar{h}) .$$

Wir definieren nun ein $\hat{h}$ durch folgende Nullbedingung:

(N) $\hat{\pi}(z,\hat{h}) = 0$ identisch in z.

Wenn ein solches $\hat{h}\in\mathbb{C}$ existiert, folgt aus (2.4) und (2.5) mit der Abkürzung $\hat{q}:=\hat{h}\beta_k$, falls $\hat{q}\neq 1$:

$$(2.6) \quad \overset{m}{\pi}(z,\hat{h}) = \frac{\pi(z,\hat{h})}{1-\hat{q}} = \pi^*(z,\hat{h}) \text{ identisch in } z.$$

Daraus folgt weiter: Ist $\hat{z}_i$ (i=1,...,k) eine Wurzel von $\pi(z,\hat{h})$, so auch von $\overset{m}{\pi}(z,\hat{h})$ und von $\pi^*(z,\hat{h})$:

$$(2.7) \quad \overset{m}{\pi}(\hat{z}_i,\hat{h}) = \pi(\hat{z}_i,\hat{h}) = \pi^*(\hat{z}_i,\hat{h}) = 0 \qquad i=1,\dots,k.$$

Die Bedingung (N) bedeutet, daß in dem Polynom $\hat{\pi}$ die Koeffizienten der Potenzen von z verschwinden. Mit den Hilfsgrößen

$$(2.8) \quad (a)\ \hat{\alpha}_i:=\alpha_i-(1-\hat{q})\alpha_i^* \qquad (b)\ \hat{\beta}_i:=\beta_i-(1-\hat{q})\beta_i^* \qquad i=0,\dots,k-1$$

ergeben sich so die Bedingungen

$$(2.9) \quad \hat{\alpha}_i=\hat{h}\,\hat{\beta}_i \qquad i=0,\dots,k-1.$$

3. Konstruktion eines Prädiktors zu vorgegebenem Korrektor

Wir denken uns einen Korrektor und eine reelle Zahl $\hat{h}$ vorgegeben und fragen nach einem Prädiktor, der die Nullbedingung (N) erfüllt ($\hat{h}$ soll reell sein, damit bei der folgenden Konstruktion die Koeffizienten des Prädiktors reell werden.). "Ordnung" bedeute im folgenden eine natürliche Zahl, die höchstens gleich der genauen Ordnung eines Verfahrens ist.

Satz 1. Zu jedem konsistenten j-Schritt-Korrektor ($j \geq 2$) mit einer Ordnung $p \geq j-1$ gibt es für jedes reelle $\hat{h} \neq 0$ genau einen k-Schritt-Prädiktor ($j \leq k \leq p+1$) der Ordnung $p^*=k-1$, der (N) genügt.

Beweis: Der Korrektor sei wie unter 2. als k-Schritt-Verfahren dargestellt. Konsistenz bedeutet:

(3.1) (a) $\rho(1) = 0$ (b) $\rho^*(1) = 0$

(3.2) (a) $\rho'(1) = \sigma(1)$ (b) $\rho^{*\prime}(1) = \sigma^*(1)$

Aus (3.1) folgt mit (2.8) und (2.9) für $\hat{h} \neq 0$

(3.3) $$\hat{\beta}_0 = -\sum_{i=1}^{k-1} \hat{\beta}_i - \beta_k .$$

Aus (3.2) folgt entsprechend

(3.4) $$\sum_{i=1}^{k-1} i\,\hat{\beta}_i = -k\beta_k .$$

Falls $k \geq 3$, erhält man aus $p \geq k-1$ und $p^* = k-1$ außerdem mit $t := 1/\hat{h}$:

(3.5) $$\sum_{i=1}^{k-1} i^j \left(1-\frac{j}{i}\,t\right) \hat{\beta}_i = -k^j\left(1-\frac{j}{k}\,t\right) \beta_k \qquad j=2,\dots,k-1 .$$

Sind reelle Zahlen $\hat{h} \neq 0, \beta_k \neq 0$ vorgegeben, so ist (3.4)/(3.5) ein System von k-1 linearen reellen Gleichungen für $\hat{\beta}_1,\dots,\hat{\beta}_{k-1}$. Das System wird etwas übersichtlicher, wenn man als Unbekannte statt $\hat{\beta}_i$ die Größen $(i\hat{\beta}_i/\beta_k)$ betrachtet. Die Koeffizientendeterminante $\Delta(t)$ ist ein Polynom in t vom Grad k-2. Für $t = 0$ ist sie eine Vandermonde-Determinante, die nicht verschwindet. Es sei a_j die j-te Zeile in der Determinante $\Delta(t)$ (j=1 entspreche der Gleichung (3.4).). Durch zeilenweise Differentiation von $\Delta(t)$ folgt

$$\dot{\Delta} = \frac{d\Delta}{dt} = \sum_{i=1}^{k-1} \Delta_i ,$$

wobei Δ_i aus Δ entsteht, indem a_i durch $\dot{a}_i$ ersetzt wird. Wegen $\dot{a}_1 = 0$ ist $\Delta_1 = 0$. Da $\dot{a}_i$ für $i=2,\dots,k-1$ eine Linearkombination von $a_1,\dots,a_{i-1}$ ist (wie man leicht nachrechnet), ist $\Delta_i = 0$ ($i=2,\dots,k-1$); also ist Δ unabhängig von t, also gleich $\Delta(0) \neq 0$.

Man kann also aus (3.4)/3.5) die Größen $\hat{\beta}_1,\dots,\hat{\beta}_{k-1}$ eindeutig berechnen. Aus (3.3) ergibt sich dann $\hat{\beta}_0$, und aus (2.9) erhält man $\hat{\alpha}_0,\dots,\hat{\alpha}_{k-1}$. (2.8) liefert dann (bei vorgegebenem Korrektor) die Koeffizienten α_i^*, β_i^* des Prädiktors.□

Satz 1 läßt für die Wahl von k nur $j \leq k \leq p+1$ zu; daraus folgt auch eine

Einschränkung für $p^*=k-1$. Als Beispiele seien folgende Korrektoren erwähnt:

a) Adams-Moulton-Verfahren (s. [5] S. 194 ff oder [6] S. 41, 85):

$p=j+1$, also $j \leq k \leq j+2$ und $p-2 \leq p^* \leq p$.

b) Backward-Differentiation-Methods (s. [6] S. 242 oder [5] S. 206 ff):

$p=j$, also $j \leq k \leq j+1$ und $p-1 \leq p^* \leq p$.

Hat man ein Prädiktor-Korrektor-Paar gefunden, das zu einem gewissen $\hat{h}$ die Bedingung (N) erfüllt, so kann man fragen, ob sie vom selben Paar auch noch für andere $\hat{h}$-Werte erfüllt wird. Hierauf gibt der folgende Satz eine Antwort ("stabil" ist hier im Sinne der Wurzelbedingung von Dahlquist zu verstehen):

Satz 2. Sind Prädiktor und Korrektor konsistent und ist der Prädiktor stabil, so hat (N) höchstens eine Lösung $\hat{h}$.

Beweis: Wir denken uns (N) für $z=1$ angeschrieben. Aus (3.2) folgt wegen der Stabilität des Prädiktors $\sigma^*(1) \neq 0$; mit (3.1) erhalten wir eine quadratische Gleichung für $\hat{h}$ mit den Lösungen

$$(3.6) \qquad \hat{h}_1 = 0, \quad \hat{h}_2 = \frac{\sigma^*(1)-\sigma(1)}{\beta_k \, \sigma^*(1)} .$$

Für $\hat{h}=\hat{h}_1$ folgt aus (N), daß ρ und ρ^* identisch sind. Dann ist aber $\rho'(1)=\rho^{*\prime}(1)$, und mit (3.2) folgt $\hat{h}_2 = 0$. Daraus folgt: Wenn (N) überhaupt eine Lösung $\hat{h} \neq 0$ hat (für beliebiges z), so ist dies $\hat{h}_2$.□

4. Absolute Stabilität

Wir bezeichnen ein Verfahren als "absolut stabil im Punkt $\bar{h}$", wenn alle Wurzeln des Stabilitätspolynoms betragsmäßig <1 sind. Die Menge aller dieser Punkte $\bar{h}$ heiße kurz "Stabilitätsbereich" des Verfahrens. Unter "PKV" sei stets ein $P(EC)^mE$-Verfahren verstanden.

Satz 3. Ist der Korrektor im Punkt $\bar{h} = \hat{h} < 0$ mit $\hat{q} \neq 1$ absolut stabil, so existiert für den gemäß Satz 1 konstruierten Prädiktor, für den Korrektor selbst und das zugehörige PKV je eine Umgebung von $\hat{h}$, in der das betreffende Verfahren absolut stabil ist.

Beweis: Wegen (2.7) sind Prädiktor und PKV im Punkt $\hat{h}$ absolut stabil. Das Polynom $\pi^*(z,\bar{h})$ ist für jedes $\bar{h}$ vom Grad k in z, hat also genau k Wurzeln $z_i(\bar{h})$. Wegen $|z_i(\hat{h})|<1$ für alle $i=1,\dots,k$ gibt es aus Stetigkeitsgründen eine Umgebung von $\hat{h}$, wo der Prädiktor absolut stabil ist. Entsprechend argumentiere man für Korrektor und PKV; jedoch ist hierbei zu berücksichtigen, daß $\overset{m}{\pi}$ gemäß (2.4) nur für $\bar{q} \neq 1$ definiert ist und π nur für $\bar{q} \neq 1$ vom Grad k bleibt. Wegen $\hat{h}\beta_k = \hat{q} \neq 1$ gibt es aber jeweils eine Umgebung von $\hat{h}$, in der $\bar{q} = \bar{h}\beta_k \neq 1$ bleibt. ¤

Auf der reellen Achse erhält man so für jedes Verfahren ein Intervall der absoluten Stabilität:

$$\bar{h}_2 < \bar{h} < \bar{h}_1 \text{ mit } \bar{h}_2 < \hat{h} < \bar{h}_1 < 0.$$

Im Hinblick auf steife Systeme ist nun wichtig, ob ein solches Verfahren außerdem in einem Intervall der Gestalt $\bar{h}_o < \bar{h} < 0$ absolut stabil ist. In der komplexen Ebene bekommt man dann insgesamt zwei Stabilitätsgebiete (die auch zu einem einzigen zusammenhängenden Gebiet verschmelzen können). Wir wollen ein derartiges Verfahren $(\hat{h}|0)$-stabil nennen. Wir benötigen folgende Verallgemeinerung von Theorem 4.6.4 in [12] S. 267:

Satz 4. Ein k-Schritt-Verfahren habe ein Stabilitätspolynom $\tilde{\pi}(z,\bar{h})$ mit $\tilde{\pi}(z,0)=\rho(z)$, wobei $\rho(1)=0$. Für die Wurzeln $z_1,\ldots,z_k$ von ρ gelte $|z_i| \leq 1$, wobei jedes z_i mit $|z_i|=1$ eine einfache Wurzel ist, deren "Wachstumsparameter"

$$(4.1) \qquad w_i = -\left[\frac{\partial\tilde{\pi}/\partial\bar{h}}{z\cdot\partial\tilde{\pi}/\partial z}\right]_{z=z_i,\ \bar{h}=0}$$

positiven Realteil hat. Dann existiert ein $\bar{h}_o < 0$, so daß für alle $\bar{h}$ mit $\bar{h}_o < \bar{h} < 0$ das Verfahren absolut stabil ist.

Beweis: Wir entwickeln die Nullstellen $\tilde{z}_i(\bar{h})$ von $\tilde{\pi}(z,\bar{h})$ in einer Umgebung von $\bar{h}=0$ mit $\tilde{z}_i(0)=z_i$ wie folgt, mit zunächst unbekannten w_i:

$$(4.2) \qquad \tilde{z}_i(\bar{h}) = z_i\cdot(1+w_i\bar{h}+O(\bar{h}^2)).$$

Falls $|z_i| < 1$, ist offensichtlich $|\tilde{z}_i(\bar{h})| < 1$ für alle $\bar{h}$ in einer gewissen Umgebung von 0. Für $|z_i| = 1$ benötigen wir w_i; durch Einsetzen von (4.2) in $\tilde{\pi}(\tilde{z}_i(\bar{h}),\bar{h}) = 0$ erhalten wir (4.1). Für Re $(w_i) > 0$ folgt dann aus (4.2) die Existenz einer negativen Zahl $\bar{h}_{o,i}$, für die gilt: $|\tilde{z}_i(\bar{h})| < 1$ für $\bar{h}_{o,i} < \bar{h} < 0$. ¤

Für den Korrektor lautet (4.1) speziell wegen (2.3)

$$(4.3) \qquad w_i = \frac{\sigma(z_i)}{z_i\rho'(z_i)}$$

(vgl. [5] S. 237 oder [12] S. 231); für das PKV erhält man aus (2.4)/(2.5):

$$(4.4) \qquad w_i = \frac{\sigma(z_i) - \beta_k \overset{m}{\rho}(z_i)}{z_i \rho'(z_i)} \quad \text{mit } \overset{m}{\rho}(z) = \begin{matrix} \rho^*(z) \\ \rho\ (z) \end{matrix} \ \text{für} \ \begin{matrix} m=1 \\ m>1 \end{matrix} .$$

Auf den Prädiktor allein ist Satz 4 mit ρ^* statt ρ anzuwenden; in (4.3) sind dann σ und ρ mit * zu versehen. Den Prädiktor können wir aber bei unserer Konstruktion nicht aussuchen, da er durch den Korrektor bestimmt wird. Man wähle also einen Korrektor so, daß er den Voraussetzungen von Satz 4 mit (4.4) $(m > 1)$ genügt; dann hat auch das $P(EC)^m E$-Verfahren für $m > 1$ das gewünschte Stabilitätsverhalten. Damit dies auch für $m = 1$ unabhängig vom Prädiktor erreicht wird, soll ρ außer $z_1=1$ nur Wurzeln mit $|z_i|<1$ besitzen. (Beachte: Für $z_1=1$ folgt aus (4.3) und (4.4) wegen (3.1)/(3.2) stets $w_1=1$.)

Unser PKV ist also $(\hat{h}|0)$-stabil, wenn der Korrektor im Punkt $\hat{h}$ absolut stabil ist und wenn die von 1 verschiedenen Wurzeln von ρ betragsmäßig < 1 sind. Soll das PKV für alle $\hat{h}<0$ $(\hat{h}|0)$-stabil sein, muß der Korrektor für alle $\bar{h} < 0$ absolut stabil sein, d. h. er muß A_0-stabil sein im Sinne von [2]. Bekanntlich haben die Backward-Differentiation-Methods für k=1, ...,6 diese Eigenschaft (vgl. [6] S. 242).

Ein $(\hat{h}|0)$-stabiles Verfahren dürfte zur Integration eines <u>steifen Systems</u> geeignet sein, das wie folgt beschaffen ist: Die Eigenwerte der Jacobi-Matrix verteilen sich auf zwei Bereiche in der linken Halbebene, nämlich (a) eine "enge" Umgebung eines (ia. variablen) Punktes Λ auf der negativen reellen Achse, (b) einen Bereich, auf dessen Rand der Punkt O liegt. Die Eigenwerte in (b) seien betragsmäßig $\ll |\Lambda|$. Λ ist somit eine Näherung für den betragsmäßig größten Eigenwert. (Für Systeme dieser Art werden auch in [8] PECE-Verfahren konstruiert.)

Um ein solches System zu integrieren, setze man $\hat{h} = h\Lambda$. Die Koeffizienten des nach Satz 1 konstruierten Prädiktors werden also (bei festem h) Funktionen von Λ , sind also evtl. variabel. Wenn die Eigenwerte sehr stark variieren, kann unser PKV versagen, da unsere Theorie eigentlich konstantes $\hat{h}$ voraussetzt. Ob ein PKV für ein gegebenes steifes System geeignet ist, kann man nur dann a priori beurteilen, wenn man den Stabilitätsbereich des Verfahrens und die Eigenwerte des Systems hinreichend gut kennt.

Der Stabilitätsbereich eines impliziten linearen Mehrschrittverfahrens bleibt bekanntlich unverändert, wenn man die implizite Formel durch <u>Newton-Iteration</u> auflöst. Bei diesem Verfahren ist es i.a. leichter als bei unserem PKV, die stabile Integration eines steifen Systems zu gewährleisten. Wenn unser PKV für ein gegebenes System überhaupt geeignet ist, sollte man es mit dem Newton-Verfahren hinsichtlich des Rechenaufwands vergleichen. Wenn das Newton-Verfahren etwa in jedem Integrationsschritt eine zeitraubende Berechnung einer Jacobi-Matrix oder eine Inversion einer großen Matrix verlangt, wird unser PKV günstiger sein (sofern Λ verhältnismäßig einfach zu bestimmen ist). Der Aufwand für die Konstruktion des Prädiktors (d. h. für die Berechnung seiner Koeffizienten als Funktionen von $\hat{h}$) wächst zwar stark mit der Schrittzahl k, hängt aber nicht von der Dimension des Systems ab. - Weitere praktische Aspekte werden am Schluß von 6. berücksichtigt.

5. Konvergenz des Stabilitätsbereichs für $m \to \infty$

Die Gestalt des in 4. untersuchten Stabilitätsbereichs des PKV lernen wir noch besser kennen, wenn wir die Wurzeln $\overset{m}{z}_i$ von $\overset{m}{\pi}(z,\bar{h})$ für $m \to \infty$ untersuchen. Zu diesem Zwecke führen wir folgende Punktmengen ein:

- S Stabilitätsbereich des Korrektors.
- S' Menge aller $\bar{h}$ mit $|z_i|>1$ für mindestens eine Nullstelle z_i von $\pi(z,\bar{h})$.
- $\overset{m}{S},\overset{m}{S}{}'$ entsprechende Mengen für $\overset{m}{\pi}$ statt π.
- J bzw. J' Menge aller $\bar{h}$ mit $|\bar{q}|<1$ bzw. >1.
- $\hat{H}$ bzw. $\hat{H}^c$ Menge aller $\bar{h}$, für die (N) (mit $\bar{h}$ statt $\hat{h}$) gilt bzw. nicht gilt.

Gemäß [11] konvergiert $\overset{m}{S}$ für $m\to\infty$ gegen S genau dann, wenn S in J enthalten ist (siehe auch [12] S. 328). Um diese Aussage zu erweitern, verwenden wir folgenden Begriff aus der Mengenlehre:

Gegeben sei eine Folge von Punktmengen $\overset{m}{A}$ $(m=1,2,\ldots)$. Die Menge $\underline{\lim}_{m\to\infty} \overset{m}{A}$ ist definiert als die Menge aller Punkte p, für die es ein $m_0(p)$ gibt, so daß alle $m \geq m_0(p)$ p Element von $\overset{m}{A}$ ist.

Wir beschreiben nun die Konvergenz von $\overset{m}{S}$, indem wir möglichst "große" Teilmengen von

$$\overset{\infty}{S} = \underline{\lim}_{m\to\infty} \overset{m}{S} \text{ und } \overset{\infty}{S}{}' = \underline{\lim}_{m\to\infty} \overset{m}{S}{}'$$

ermitteln. Wegen (2.7) gilt für jedes m

$$(5.1) \qquad (a) \quad \overset{m}{S} \cap \hat{H} = S \cap \hat{H} \qquad (b) \quad \overset{m}{S}{}' \cap \hat{H} = S' \cap \hat{H} \;;$$

daraus folgt

$$(5.2) \qquad (a) \quad S \cap \hat{H} \subset \overset{\infty}{S} \qquad (b) \quad S' \cap \hat{H} \subset \overset{\infty}{S}{}' \;.$$

Für $\bar{h} \in J$ folgt aus (2.4)

$$(5.3) \qquad \lim_{m\to\infty} \overset{m}{\pi}(z,\bar{h}) = \frac{\pi(z,\bar{h})}{1-\bar{q}}$$

Da in den Polynomen $\overset{m}{\pi}(z,\bar{h})$ der Koeffizient von z^k stets 1 ist, hängen ihre Nullstellen $\overset{m}{z}_j$ stetig von den übrigen Koeffizienten ab. Die $\overset{m}{z}_j$ liegen also wegen (5.3) beliebig nahe an den Nullstellen z_i von $\pi(z,\bar{h})$, wenn nur m hinreichend groß ist; d. h. in jeder Umgebung von z_i liegt für hinreichend großes m mindestens ein $\overset{m}{z}_j$. Daraus folgt aber, da wir $\bar{h} \in J$ vorausgesetzt haben:

$$(5.4) \qquad (a) \quad S \cap J \subset \overset{\infty}{S} \qquad\qquad (b) \quad S' \cap J \subset \overset{\infty}{S}' .$$

Aus (5.2a) und (5.4a) erhalten wir:

$$(5.5) \qquad S \cap (\hat{H} \cup J) \subset \overset{\infty}{S} .$$

Um eine entsprechende Teilmenge von $\overset{\infty}{S}'$ zu bekommen, betrachten wir noch den Fall $\bar{h} \in \hat{H}^c \cap J'$. Wir formen (2.4) um:

$$(5.6) \qquad \overset{m}{\pi}(z,\bar{h}) = \frac{\bar{q}^{m}}{1-\bar{q}} \left[\pi(z,\bar{h})\, \bar{q}^{-m} + \hat{\pi}(z,\bar{h}) \right]$$

In den eckigen Klammern steht ein Polynom in z vom Grad k. Der Koeffizient von z^k konvergiert für $m \to \infty$ gegen 0, während mindestens einer der übrigen Koeffizienten wegen $\bar{h} \in \hat{H}^c$ gegen einen von 0 verschiedenen Wert strebt. Deshalb gibt es zu jeder Zahl $M > 0$ ein $m_o > 0$, so daß für alle $m \geq m_o$ mindestens eine Wurzel z_i von $\overset{m}{\pi}$ betragsmäßig M übertrifft (vgl. etwa [10]). Daraus folgt:

$$(5.7) \qquad \hat{H}^c \cap J' \subset \overset{\infty}{S}'$$

Wir fassen (5.7) mit (5.2b) und (5.4b) zusammen zu

$$(5.8) \qquad (S' \cap (\hat{H} \cup J)) \cup (\hat{H}^c \cap J') \subset \overset{\infty}{S}' .$$

Um die Ergebnisse (5.5) und (5.8) zu interpretieren, spezialisieren wir sie zunächst auf den Fall, daß $\hat{H}$ leer ist:

$$(5.9) \qquad (a) \quad S \cap J \subset \overset{\infty}{S} \qquad\qquad (b) \quad (S' \cap J) \cup J' \subset \overset{\infty}{S}' .$$

So "konvergieren" die Stabilitätsbereiche $\overset{m}{S}$ gegen den nur vom Korrektor abhängigen Bereich $S \cap J$. Dieser Bereich schneidet auf der reellen Achse meist ein Intervall $\bar{h}_\infty < \bar{h} < 0$ aus; als Beispiel mögen wieder folgende Korrektoren dienen:

a) <u>Adams-Moulton-Verfahren:</u> Wie man aus [4] durch einige Zwischenrechnungen unter Verwendung von [5] S. 192 ff herleiten kann, gehört das betragsmäßig größte $\bar{h}_\infty$ zu k=3, nämlich $\bar{h}_\infty=-8/3$.

b) <u>Backward-Differentiation-Methods:</u> Beschränkt man sich auf die stabilen Verfahren, nämlich mit $1 \leq k \leq 6$ (s. [1]), so erhält man den größten Wert von $|\bar{h}_\infty|$ für k=6 : $\bar{h}_\infty= -2.45$ (zum Beweis verwende man die A_o-Stabilität dieser Verfahren; außerdem ergibt [5] S. 207 f für J die Menge $|\bar{h}|< \sum_{i=1}^{k} \frac{1}{i}$.)

Ist $\hat{H}$ nicht leer, so ist für uns vor allem der Fall wichtig, daß ein $\hat{h} \in \hat{H}$ zwar in S, aber nicht in J liegt (d. h. $|\hat{q}|>1$). Dann enthält $\overset{\infty}{S}$ wegen (5.5) den außerhalb von $S \cap J$ gelegenen Punkt $\hat{h}$. Man kann also grob sagen: Für großes m besteht $\overset{m}{S}$ nahezu aus $S \cap J$ und einer kleinen Umgebung von $\hat{h}$. Da diese Umgebung für $m \to \infty$ auf den Punkt $\hat{h}$ zusammenschrumpft, sollte man m nicht sehr groß werden lassen.

6. Lokaler Verfahrensfehler

Der lokale Verfahrensfehler (lVf) unseres PKV soll hier nur für die skalare Differentialgleichung

$$(6.1) \qquad y' = \lambda y + b(x)$$

mit einer (i.a. komplexen) Konstanten $\lambda \neq 0$ diskutiert werden. Daraus ergeben sich wenigstens gewisse Anhaltspunkte für den lVf bei Anwendung auf ein (lineares oder nichtlineares) System.

Das PKV hat die Gestalt eines nichtlinearen Mehrschrittverfahrens, wie es etwa in [13] S. 121ff zu finden ist:

$$(6.2) \qquad \sum_{i=0}^{k} \alpha_i y_{n+i} = h\ F(x_n;\ y_{n+k},\ y_{n+k-1},\dots,\ y_n;\ h)$$

mit $\alpha_k=1$. Der Korrektor ist ebenfalls ein spezielles Verfahren (6.2), ebenso der Prädiktor (wenn α_i durch α_i^* ersetzt wird).

Durch einen Punkt (x,y) sei genau eine Lösung v(x) der zu integrierenden Gleichung (d. h. hier (6.1)) bestimmt. Wir bezeichnen dann als lVf:

$$(6.3) \qquad \bar{T}(x,y,h) = \sum_{i=0}^{k} \alpha_i v(x+ih)-hF(x;v(x+kh),\dots,v(x)\ ;h).$$

Die Argumente (x,y,h) lassen wir im folgenden weg, da kein Mißverständnis zu befürchten ist. Im Fall des Korrektors bzw. des PKV schreiben wir für $\bar{T}$ stets T bzw. $\overset{m}{T}$; entsprechend bedeutet T^* (mit α_i^* statt α_i) den lVf des Prädiktors. Hiermit ergibt sich bei Anwendung auf (6.1) nach einer Zwischenrechnung:

$$\overset{m}{T} = \bar{q}^m T^* + \frac{1-\bar{q}^m}{1-\bar{q}} T . \tag{6.4}$$

Diese Beziehung diskutieren wir auf zwei Arten:

I. Wir nehmen an, daß (N) eine Lösung $\hat{h}$ besitzt, und gehen von einer Schrittweite h in der Nähe von $\hat{h}/\lambda$ aus (d. h. $\bar{h}\approx\hat{h}$). Wir spalten b(x) auf in einen konstanten Anteil $\bar{b}$ und eine Abweichung $\delta(x)$: $b(x) = \bar{b}+\delta(x)$. Dann ergibt (6.4) (zusammen mit der Definition von T^* und T) unter Verwendung von (2.6), (2.8), (2.9) und (3.1), wobei noch $\hat{\beta}_k=\beta_k$ gesetzt sei:

$$\overset{m}{T} = \frac{T}{1-\bar{q}} + \frac{\bar{q}^m}{1-\hat{q}} \left[\frac{\hat{q}-\bar{q}}{1-\bar{q}} T - D \right] \quad \text{mit} \tag{6.5}$$

$$D = \sum_{i=0}^{k} \hat{\beta}_i \cdot \left((\hat{h}-\bar{h})\left(v(x+ih) + \frac{\bar{b}}{\lambda}\right) - \frac{\bar{h}}{\lambda}\delta(x+ih)\right) . \tag{6.6}$$

Im "Idealfall" $\bar{h}=\hat{h}$, $\delta(x)=0$ ist D=0 und $\overset{m}{T}= \frac{T}{1-\hat{q}} = T^*$ (analog zu (2.6)). Andernfalls ist in (6.5) der Inhalt von [ ] i.a. $\neq 0$. In dem für uns interessanten Fall $|\bar{q}|>1$ (vgl. den Schluß von 5.) wird aber [ ] mit einem Faktor multipliziert, dessen Betrag für $m\to\infty$ unbeschränkt anwächst. Deshalb sollte man m möglichst klein halten, und wir werden uns im folgenden vor allem mit dem PECE-Verfahren beschäftigen.

II. Wir untersuchen das asymptotische Verhalten für $h \to 0$. Für Korrektor und Prädiktor gibt es die bekannten Formeln

$$\begin{aligned} T &= C_{p+1} v^{(p+1)}(x) h^{p+1} + O(h^{p+2}) \\ T^* &= C^*_{p^*+1} v^{(p^*+1)}(x) h^{p^*+1} + O(h^{p^*+2}) \end{aligned} \tag{6.7}$$

mit den Fehlerkonstanten C_{p+1}, $C^*_{p^*+1}$. Das asymptotische Verhalten bei einem PKV wurde in [6] S. 88ff dargestellt; wir beschränken uns jetzt auf das PECE-Verfahren. Ist $p^* \geq p$, so gilt für $\overset{1}{T}$ dieselbe Darstellung wie für T in (6.7). Ist $p^* \leq p-2$, so wird $\overset{1}{T} = O(h^{p^*+2})$, d. h. um mindestens eine Ordnung schlechter als T. Für $p^*=p-1$ wird $\overset{1}{T} = O(h^{p+1})$, und zwar erhält man für unsere Testgleichung (6.1) aus (6.4) und (6.7):

$$\overset{1}{T}=[\tilde{C}_p \cdot (\lambda^{p+1} v(x) + \lambda b_p(x)) + C_{p+1} b^{(p)}(x)]\; h^{p+1} + O(h^{p+2}) \tag{6.8}$$

mit $\tilde{C}_p = \beta_k C_p^* + C_{p+1}$ und $b_p(x) = \sum_{i=0}^{p-1} \lambda^i b^{(p-i-1)}(x)$.

Sonderfall $b^{(p)}(x)=0$: (6.8) vereinfacht sich zu

$$\text{(6.9)} \qquad \overset{1}{T} = \tilde{C}_p v^{(p+1)}(x) h^{p+1} + O(h^{p+2}) .$$

Wenn es gelingt, $\hat{h}$ so zu wählen, daß $|\tilde{C}_p| < |C_{p+1}|$ wird, so wird also das PECE-Verfahren genauer sein als der Korrektor; falls $\tilde{C}_p = 0$ erreicht wird, ist $\overset{1}{T}$ sogar um eine Ordnung besser als T. Diese Möglichkeit ist aber nur dann in Betracht zu ziehen, wenn $\hat{h}$ nicht anderweitig (nämlich durch Stabilitätsforderungen) bestimmt wird. Unsere Konstruktion in Satz 1 läßt sich so auch für nichtsteife Systeme ausnützen.

Bei den in 4. beschriebenen Systemen treten entsprechend (a) und (b) die Fälle I und II gleichzeitig auf. Praktische Folgerung für steife Systeme:

Im Hinblick auf I ist das PECE-Verfahren besonders günstig. Bei der Anwendung von II auf ein steifes System ist zu beachten, daß der Term $O(h^{p+2})$ in (6.7), (6.8) oder (6.9) mit der aktuellen Schrittweite h evtl. gegenüber dem Term der Ordnung $O(h^{p+1})$ dominiert. Ist dies nicht der Fall, sollte (gemäß obigen Ausführungen) $p^* \geq p-1$ gewählt werden. (Im Anschluß an Satz 1 wurden ebenfalls Schranken für p^* genannt.) Damit der Rechenaufwand für Prädiktor und PKV nicht unnötig groß wird, wähle man die Schrittzahl k möglichst klein, also am besten so, daß $p^*=p-1$ wird, d. h. k=p.

7. Verfahren mit höheren Ableitungen

Die bisherigen Ausführungen gelten zum größten Teil auch, wenn bei Prädiktor und Korrektor höhere Ableitungen verwendet werden (Obrechkoff-Verfahren, s. [6] S. 199ff oder [12] S. 347ff; PKV mit höheren Ableitungen werden ausführlicher in [9] S. 45ff behandelt). In diesem Abschnitt wird nur besprochen, welche Änderungen nötig sind. Es sollen Ableitungen bis zur Ordnung r auftreten.

In (2.1) bzw. (2.2) ersetze man σ bzw. σ^* durch

$$\text{(7.1)} \qquad \sigma_n(z) = \sum_{i=0}^{k} \beta_{in} z^i \quad \text{bzw.} \quad \sigma_n^*(z) = \sum_{i=0}^{k-1} \beta_{in}^* z^i \qquad n=1,\dots,r .$$

Die Stabilitätspolynome lauten statt (2.3):

(7.2) $\pi(z,\bar{h}) = \rho(z) - \sum_{n=1}^{r} \bar{h}^n \sigma_n(z), \qquad \pi^*(z,\bar{h}) = \rho^*(z) - \sum_{n=1}^{r} \bar{h}^n \sigma_n^*(z).$

Um wieder auf die Darstellung (2.4)/(2.5) sowie auf (2.6) zu kommen, setze man

(7.3) $\bar{q} := \sum_{n=1}^{r} \bar{h}^n \beta_{kn}$ und $\hat{q} := \sum_{n=1}^{r} \hat{h}^n \beta_{kn}$

Die Hilfsgrößen $\hat{\beta}_i$ aus (2.8) ersetze man durch

(7.4) $\hat{\beta}_{in} = \beta_{in} - (1-\hat{q})\beta^*_{in} \qquad i=0,\dots,k-1; n=1,\dots,r.$

Die Nullbedingung (N) liefert jetzt an Stelle von (2.9):

(7.5) $\hat{\alpha}_i = \sum_{n=1}^{r} \hat{h}^n \hat{\beta}_{in} \qquad i=0,\dots,k-1.$

Ob Satz 1 allgemein übertragen werden kann, ist noch nicht bekannt. Man kann aber ein lineares Gleichungssystem für die $\hat{\beta}_{in}$ aufstellen, falls $p \geq rj-1$, $j \leq k \leq \frac{p+1}{r}$ und $p^*=rk-1$. Hierzu ersetze man (3.2) durch

(7.6) (a) $\rho'(1) = \sigma_1(1)$ (b) $\rho^{*\prime}(1) = \sigma_1^*(1)$.

Die Bedingungen für die Koeffizienten eines Obrechkoff-Verfahrens, das eine bestimmte Ordnung haben soll, findet man in [3] . Aus diesen Bedingungen (die auch die Konsistenzbedingungen (3.1) und (7.6) umfassen) kann man für $k \geq 2$ ein kompliziertes System von $(k-1)r$ linearen Gleichungen für $\hat{\beta}_{in}$ $(i=1,\dots,k-1; n=1,\dots,r)$ herleiten. Über die Lösbarkeit dieses Systems ist noch keine allgemeine Aussage bekannt. Falls die $\hat{\beta}_{in}$ hieraus bestimmt werden können, lassen sich die $\hat{\beta}_{on}$ $(n=1,\dots,r)$ ebenfalls ermitteln (aus einer Formel, die bei der Herleitung des genannten Systems erscheint). Aus (7.4) und (2.8a) mit (7.5) berechne man dann die Koeffizienten des Prädiktors.

Im Fall $k=1$ dagegen läßt sich der Prädiktor explizit angeben. Er hat die Ordnung $r-1$ genau dann, wenn $\alpha_1^* = -\alpha_o^* = 1$ und $\beta^*_{on} = \frac{1}{n!}$ $(n=1,\dots,r-1)$. Der noch fehlende Koeffizient β^*_{or} ergibt sich aus (7.5) mit (2.8a) und (7.4), wenn der Korrektor als konsistent vorausgesetzt wird:

$$\beta^*_{or} = \frac{1}{\hat{h}^r} \sum_{n=1}^{r-1} \hat{h}^n \left(\frac{\beta_{on}+\beta_{1n}}{1-\hat{q}} - \frac{1}{n!} \right) + \frac{\beta_{or}+\beta_{1r}}{1-\hat{q}}$$

Satz 2 konnte noch nicht verallgemeinert werden. Satz 3 bleibt samt Beweis unverändert, ebenso Satz 4. In (4.3) und (4.4) ersetze man σ durch σ_1 sowie β_k durch β_{k1}. Die allgemeinen Ausführungen unter 5. bleiben gültig. In 6. muß man β_k durch β_{k1} ersetzen und den Ausdruck (6.6) für D verallgemeinern.

Als Beispiele für Korrektoren seien jetzt zwei Klassen von Einschrittverfahren genannt (hierbei setzen wir k=1; falls ein Prädiktor mit $k>1$ dazukonstruiert werden soll, muß man die angegebenen Koeffizienten α,β umnumerieren):

a) Hermite-Verfahren: Hiermit bezeichnen wir die auf der Hermite-Interpolation beruhenden Verfahren mit $\alpha_o=-1$ und

$$\beta_{on}=-(-1)^n\beta_{1n}=C_{nr}:=\frac{1}{n!}\cdot\frac{r(r-1)\ldots(r-n+1)}{2r(2r-1)\ldots(2r-n+1)}\ .$$

Sie haben die Ordnung 2r und sind A-stabil, s. [3] oder [7] (ausführlicher behandelt in [9] S. 83ff). Konstruiert man hierzu einen Prädiktor mit k=1 bzw. k=2, so erhält dieser die Ordnung $p^*=r-1$ (vgl. obige Konstruktion im Fall k=1) bzw. $p^*=2r-1$. ($k>2$ ist für $r>1$ nicht zulässig.) Die unter 5. eingeführte "linke" Grenze von $S\cap J$ werde mit $\bar{h}_{\infty,r}$ bezeichnet; man kann leicht beweisen:

$$-2=\bar{h}_{\infty,1}<\bar{h}_{\infty,2}<\bar{h}_{\infty,3}<\ldots<-2\ln 2\approx-1.3863.$$

b) Taylor-Verfahren: Durch Taylorentwicklung kann man das implizite Verfahren mit $\alpha_o=-1$, $\beta_{on}=0$ und $\beta_{1n}=-\frac{(-1)^n}{n!}$ herleiten, das die Ordnung r hat und A_o-stabil ist. Der Prädiktor mit k=1 erhält die Ordnung $p^*=r-1$; für $r>1$ ist $k>1$ verboten. Man zeigt leicht:

$$-1=\bar{h}_{\infty,1}<\bar{h}_{\infty,2}<h_{\infty,3}<\ldots<-\ln 2\approx-0.69315.$$

8. Beispiel

Die bekannte Trapezregel (p=2) ist sowohl ein Adams-Moulton- als auch ein Hermite-Verfahren. Der zugehörige Zweischritt-Prädiktor mit $p^*=1$ hat die Koeffizienten

$$\beta_o^*=\frac{1}{\hat{h}-2}\qquad\beta_1^*=-3\beta_o^*\qquad\alpha_o^*=\hat{h}\beta_o^*\qquad\alpha_1^*=2(1-\hat{h})\beta_o^*\ .$$

ρ^* hat die Nullstellen $z_1=1$, $z_2=\frac{\hat{h}}{\hat{h}-2}$. Für $\hat{h}<0$ (was stets vorausgesetzt sei) ist $|z_2|<1$, d. h. der Prädiktor ist stabil und (nach den Sätzen 3 und 4) $(\hat{h}|0)$-stabil. Man kann sogar zeigen, daß der Prädiktor für alle $\bar{h} \in (\hat{h}-1,0)$ absolut stabil ist (d. h. beide "Teile" des Stabilitätsbereichs hängen zusammen).

Das PECE-Verfahren hat ebenfalls einen zusammenhängenden Stabilitätsbereich, sofern $\hat{h} > \hat{\hat{h}} = -1-2\sqrt{2} \approx -3.83$ bleibt, und zwar ist es dann absolut stabil für alle $\bar{h} \in (\bar{h}_o, 0)$ mit einem gewissen $\bar{h}_o \in (\hat{h}-2,\hat{h})$. Für $\hat{h} \le \hat{\hat{h}}$ ist das Verfahren absolut stabil für alle $\bar{h} \in (\bar{h}_o, 0)$ und $\bar{h} \in (\bar{h}_2, \bar{h}_1)$ mit einem gewissen $\bar{h}_2 \in (\hat{h}-2,\hat{h})$ und

(8.1) $$\bar{h}_{o,1} = -\frac{1-\hat{h}}{2} \pm \frac{1}{2}\sqrt{(1+\hat{h})^2-8}\ .$$

Aus (8.1) folgt für $\hat{h} \to -\infty$

(8.2) $$\bar{h}_o \sim -1-\frac{2}{-1-\hat{h}},\quad \bar{h}_1 \sim \hat{h} + \frac{2}{-1-\hat{h}}\ .$$

Bei beliebigem $\hat{h}<0$ enthält der Stabilitätsbereich das Intervall $-1 \le \bar{h} < 0$.

Die Trapezregel hat die Fehlerkonstante $C_3 = -\frac{1}{12}$, der Prädiktor: $C_2^* = \frac{\hat{h}}{\hat{h}-2}$. Die unter 6. ("Sonderfall") gewünschte Bedingung $\tilde{C}_2 = 0$ wird daher für $\hat{h} = -\frac{2}{5}$ erfüllt.

Als Beispiel für ein steifes System haben wir ein nichtlineares, autonomes System von 9 Gleichungen gewählt:

(8.3) $$\begin{aligned} y_i' &= -\sum_{n=1}^{i} n y_n + \frac{i}{10} y_9 \qquad i=1,\dots,8 \\ y_9' &= (s-\lambda\cdot(1-y_9^2))y_9 \quad \text{mit} \quad s = \sum_{n=1}^{8} y_n^2 \quad \text{und } \lambda < 0\,. \end{aligned}$$

Die Jacobi-Matrix hat angenähert die Eigenwerte $-1,-2,\dots,-8$ und

(8.4) $$\Lambda = s - \lambda\cdot(1-3y_9^2).$$

Die einzigen konstanten nichttrivialen Lösungen von (8.3) sind gegeben durch

(8.5) $$\bar{y}_i = \frac{1}{10i}\,\bar{y}_9 \quad (i=1,\dots,8) \text{ mit } \bar{y}_9 = \pm\left(1+\frac{1}{100\lambda}\sum_{n=1}^{8}\frac{1}{n^2}\right)^{-\frac{1}{2}}$$

Aus (8.4) folgt aber für $y_9 = \bar{y}_9$: $\Lambda = 2\lambda\bar{y}_9^2 < 0$; also bezeichnet (8.5) zwei asymptotisch stabile Ruhelagen. Für "großes" $|\lambda|$ ist (8.3) deshalb in einer gewissen Umgebung jeder dieser Ruhelagen ein steifes System, wie es am Schluß von 4. beschrieben wurde.

Wir haben (8.3) mit den Anfangswerten $y_i(0)=0$ $(i=1,\ldots,8)$ und $y_9(0)=1$ für $0\leq x\leq 0.4$ mit $h = 0.01$ integriert, und zwar für $\lambda=-10^3$ und -10^4. (Der Anfangswert für y_9 liegt nahe an dem positiven Wert von $\bar{y}_9$; deshalb wird Λ im Integrationsintervall nicht zu stark schwanken.) Wir haben u. a. folgende Verfahren ausprobiert:

1) den obigen Prädiktor mit $\hat{h} = h\Lambda$,

2) das zugehörige PECE-Verfahren,

3) die Trapezregel mit einem Newton-Schritt, für den das Euler-Cauchy-Verfahren die Startwerte lieferte.

Alle diese Verfahren blieben numerisch stabil. Um die Genauigkeit der Lösungen beurteilen zu können, haben wir mit dem klassischen Runge-Kutta-Verfahren 4. Ordnung mit hinreichend kleiner Schrittweite eine Vergleichslösung berechnet. Die so ermittelte Genauigkeit für 2) lag in derselben Größenordnung wie bei 3) (teils besser, teils schlechter), während die Ergebnisse von 1) so gut wie unbrauchbar waren. (Die Schrittweite $h=0.01$ war so gewählt, daß mit 3) eine "vernünftige" Genauigkeit erzielt wurde; ein zweiter Newton-Schritt brachte kaum eine Verbesserung.) Die Rechenzeiten der Verfahren 1) - 3) verhielten sich etwa wie 2 : 3 : 16. Mit unserem PECE-Verfahren wurde also die Rechenzeit gegenüber dem Verfahren mit Newton-Iteration auf knapp ein Fünftel reduziert.

Die Verfahren 1) und 2) haben wir auch mit dem oben erwähnten Wert $\hat{h}=-\frac{2}{5}$ ausprobiert; zum Vergleich haben wir 3) verwendet, aber mit 2 Newton-Schritten. Als nichtsteifes Anwendungsbeispiel diente (8.3) mit $\lambda=-1$ und -10, wobei jetzt $y_9(0)=.75$ gewählt wurde. Die Ergebnisse von 2) waren erwartungsgemäß genauer als diejenigen von 3), und zwar durchschnittlich um mindestens eine Zehnerpotenz. Im Vergleich zu 1) war 2) um 2 - 3 Zehnerpotenzen genauer, und zwar bei höchstens doppelter Rechenzeit.

Die Rechnungen wurden auf der Rechenanlage TR440 des Rechenzentrums der Ruhr-Universität Bochum durchgeführt. Die Programme sind in FORTRAN (einfache Genauigkeit) geschrieben; für die bei 3) erforderliche Auflösung eines linearen Gleichungssystems wurde ein Bibliotheksprogramm für Gauß-Elimination verwendet.

Literatur:

[1] *Creedon, D.M./Miller, J.J.H.:* The stability properties of q-step backward difference schemes. BIT 15 (1975), S. 244-249.

[2] *Cryer, W.W.:* A new class of highly-stable methods: A_o-stable methods. BIT 13 (1973), S. 153-159.

[3] *Griepentrog, E.:* Mehrschrittverfahren zur numerischen Integration von gewöhnlichen Differentialgleichungssystemen und asymptotische Exaktheit. Wiss. Z. Humboldt Univ. Berlin Math.-Natur.-Reihe, Jahrg. 19 (1970), S. 637-653.

[4] *Hall, G.:* Stability analysis of predictor-corrector algorithms of Adams type. SIAM J. Numer. Anal. 11 (1974), S. 494-505.

[5] *Henrici, P.:* Discrete variable methods in ordinary differential equations. New York/London/Sydney: John Wiley 1962.

[6] *Lambert, J.D.:* Computational methods in ordinary differential equations. London/New York/Sydney/Toronto: John Wiley 1973.

[7] *Loscalzo, F.R.:* An introduction to the application of spline functions to initial value problems. In: *Greville, T.N.E.* (Hrsg.): Theory and applications of spline functions, New York/London: Academic Press 1969, S. 37-64.

[8] *Oesterhelt, G.:* Mehrschrittverfahren zur numerischen Integration von Differentialgleichungssystemen mit stark verschiedenen Zeitkonstanten. Computing 13 (1974), S. 279-298.

[9] *Piehler, G.:* Theorie der Obrechkoff-Verfahren und die Behandlung von Anfangswertproblemen durch Spline-Interpolation nach Loscalzo. Diplomarbeit Bochum 1974.

[10] *Rodabaugh, D.J.:* On stable correctors. Comp. J. 13 (1970), S. 98-100.

[11] *Stetter, H.J.:* Improved absolute stability of predictor-corrector schemes. Computing 3 (1968), S. 286-296.

[12] *Stetter, H.J.:* Analysis of discretization methods for ordinary differential equations. Berlin/Heidelberg/New York: Springer 1973.

[13] *Stoer, J./Bulirsch, R.:* Einführung in die Numerische Mathematik II. Berlin/Heidelberg/New York: Springer 1973.

OSCILLATION AND NONOSCILLATION THEOREMS FOR A SECOND ORDER NONLINEAR FUNCTIONAL DIFFERENTIAL EQUATION

B. MEHRI

In this note, we are concerned with the following non-linear delay differential equation

$$(r(t)\ x'(t))' + f(t,x_t) = o\ . \tag{1}$$

Before we state our main theorems, we make the following remarks. Equation (1) is considered to be a retarted equation with a maximum delay $h>o$. If $X(.)$ is defined on $[t_o-h,\infty)$, $t_o\geq o$, then for any fixed t, $t\geq t_o$, the symbol x_t will denote the restriction of $X(.)$ to the interval $[t-h,t]$, and $x_t(s)=x(t+s), s\varepsilon|-h,o|$. t_o is the initial point and the initial condition is defined on $[t_o-h,\ t_o]$. We shall restrict our attention to those solutions of (1) which exist on some ray $[t_o-h,\infty)$, where t_o may depend on the particular solution, and are non-trivial in any neighborhood of infinity. Such solution is said to be oscillatory if it has arbitrary large zeros; otherwise, it is said to be nonoscillatory. Equation (1) is said to be oscillatory if all of its solutions are oscillatory.

Throughout this paper, we shall assume:

i) $r(t)$ is continuous and positive for $t\geq t_o-h$

ii) $f(t,y)$ is continuous for $[t_o-h,\infty)\times R$ with values in R.

iii) $f(t,y)$ sign $y>o$, for each $t\varepsilon[t_o-h,\infty)$

iv) $|f(t,y_1)| \leq |f(t,y_2)|$ if $|y_1| \leq |y_2|$

or

iv)' $|f(t,y_1)| \leq |f(t,y_2)|$ if $|y_1| \leq |y_2|$.

Note. Equation (1) is called superlinear or sublinear if condition (iv) or (iv)', respectively, is satisfied.

We also define

$$p(t) = \begin{cases} \int_{t_o}^{t} \frac{d\tau}{r(\tau)} & \text{if} \quad \int^{\infty} \frac{d\tau}{r(\tau)} \quad \text{diverges} \\ \int_{t}^{\infty} \frac{d\tau}{r(\tau)} & \text{if} \quad \int^{\infty} \frac{d\tau}{r(\tau)} \quad \text{converges.} \end{cases}$$

Theorem 1. Let equation (1) be either superlinear or sublinear. In order for (1) to be oscillatory it is necessarry that for any non-zero constant c the following conditions hold

$$\int_{t_o}^{\infty} p(\tau)\ |f(\tau,c)|\ d\tau = \infty, \qquad \int_{t_o}^{\infty} |f(\tau,cp(\tau+s))|\ d\tau = \infty \ , \tag{2}$$

$s \in [-h,o]$.

Proof. In order to prove this theorem, we must show that if the conditions (2) are not satisfied, i.e. for some nonzero constant c the condition

$$\int_{t_o}^{\infty} p(\tau)\ |f(\tau,c)|\ d\tau < \infty \tag{3}$$

or

$$\int_{t_o}^{\infty} |f(\tau,cp(\tau+s))|\ d\tau < \infty \ , \tag{4}$$

$s \in [-h,o]$,

is satisfied, then there exists at least one non-oscillatory solution for (1). First we assume condition (3) holds. Then we consider the following integral equations

$$x(t) = \begin{cases} \frac{c}{2} + p(t) \int_{t}^{\infty} f(\tau,x_\tau)d\tau + \int_{t_1}^{t} p(\tau)f(\tau,x_\tau)d\tau \ ; & \text{if } p(t) \text{ diverges} \\ \frac{c}{2} + \int_{t}^{\infty} p(\tau)f(\tau,x_\tau)d\tau + \int_{t_1}^{t} f(\tau,x_\tau)\ d\tau \ ; & \text{if } p(t) \text{ converges} \end{cases} \tag{5}$$

where t_1 is chosen so that

$$\int_{t_1}^{\infty} p(\tau)|f(\tau,c)|d\tau \le \frac{|c|}{2} \ .$$

If one shows that the above integral equation possesses a continuous solution which is uniformly bounded away from zero, then x(t) will be a nonoscillatory solution of (1).
Next, we define a sequence of functions $\{x_n(t)\}$, $t \geq t_1-h$, $n = 0, 1, ..$ by

$$x_o(t)=\begin{cases} \frac{c}{2}, \ t \geq t_1-h & \text{if (1) is superlinear} \\ c, \ t \geq t_1-h & \text{if (1) is sublinear} \end{cases},$$

$$x_n(t)=\begin{cases} \frac{c}{2}+p(t)\int_t^\infty f(\tau,(x_{n-1})_\tau)\,d\tau + \int_{t_1}^t p(\tau)f(\tau,(x_{n-1})_\tau)\,d\tau & t_1 \leq t \leq \infty \\ \frac{c}{2}+\int_t^\infty p(\tau)f(\tau,(x_{n-1})_\tau)\,d\tau + p(t)\int_{t_1}^t f(\tau,(x_{n-1})_\tau)d\tau & t_1 \leq t \leq \infty \end{cases} \tag{6}$$

$$x_n(t) = x_n(t_1) \qquad t_1-h<t<t_1 .$$

Using the assumptions (i) - (iv) or (iv)', one can show by induction

$$\frac{|c|}{2} \leq x_n(t) \text{ sign } c \leq |c| \ ; t \geq t_1-h, \ n=0,1,2,... \tag{7}$$

and

$$|x_n(t)| \leq |x_{n+1}(t)| \qquad t \geq t_1-h, \ n=0,1,2,...$$

or

$$|x_{n+1}(t)| \leq |x_n(t)| \qquad t \geq t_1-h, \ n=0,1,2,... \ .$$

It follows that the sequence $\{x_n(t)\}$ converges to some function x(t). Furthermore

$$|c|/_2 \leq x(t) \text{ sign } c \leq |c|; \quad t \geq t_1-h. \tag{8}$$

We now show x(t) is a solution of the integral equation (5). For any preassigned $\varepsilon > 0$, we choose T such that

$$\int_T^\infty p(\tau) \ |f(\tau,c)| \ d\tau < \varepsilon/2.$$

Then for x(t) defined as in (5) we have

$$\begin{cases} |x_n(t)-\frac{c}{2}-p(t)\int_t^{\infty} f(\tau,x_\tau)\, d\tau - \int_{t_1}^{t} p(\tau)f(\tau,x_\tau)\, d\tau| \\ |x_n(t)-\frac{c}{2}-\int_t^{\infty} p(\tau)f(\tau,x_\tau)\, d\tau - p(t)\int_{t_1}^{t} f(\tau,x_\tau)\, d\tau| \end{cases}$$

$$\leq \int_{t_1}^{T} p(\tau)|f(\tau,(x_{n-1})_\tau)-f(\tau,x_\tau)|d\tau+\int_T^{\infty} p(\tau)|f(\tau,(x_{n-1})_\tau)-f(\tau,x_\tau)|\, d\tau$$

$$\leq \varepsilon+\int_{t_1}^{T} p(\tau)|f(\tau,(x_{n-1})_\tau) - f(\tau,x_\tau)|d\tau \quad .$$

If in the latter inequality we pass to the limit as $n \to \infty$, we obtain

$$|x(t)-\frac{c}{2}-p(t)\int_t^{\infty} f(\tau,x_\tau)d\tau-\int_{t_1}^{t} p(\tau)f(\tau,x_\tau)d\tau|<\varepsilon$$

or

$$|x(t)-\frac{c}{2}-\int_t^{\infty} p(\tau)f(\tau,x_\tau)d\tau-p(t)\int_{t_1}^{t} f(\tau,x_\tau)d\tau|<\varepsilon \quad .$$

But since ε is arbitrary, it follows from the last inequality that x(t) is a solution of the integral equation (5) as well as, as is easily checked, a solution of the differential equation (1), but on the otherhand according to (8), x(t) is nonoscillatory.
Now, assume condition (4) is satisfied, we consider the following integral equation

$$x(t) = \begin{cases} \frac{c}{2}p(t)+p(t)\int_t^{\infty} f(\tau,x_\tau)d\tau+\int_{t_1}^{t} p(\tau)f(\tau,x_\tau)d\tau \;; & \text{if } p(t) \text{ diverges} \\ \frac{c}{2}p(t)+\int_t^{\infty} p(\tau)f(\tau,x_\tau)d\tau+p(t)\int_{t_1}^{t} f(\tau,x_\tau)d\tau \;; & \text{if } p(t) \text{ converges .} \end{cases} \tag{9}$$

Case I. p(t) diverges

We define $\{x_n(t)\}_{n=1}^{\infty}$ as

$$x_0(t) = \begin{cases} \frac{c}{2}\, p(t) & \text{if (1) is superlinear} \quad t \geq t_1 - h \\ \\ c\, p(t) & \text{if (1) is sublinear} \quad t \geq t_1 - h\,, \end{cases}$$

$$x_n(t) = \tfrac{c}{2}p(t) + p(t)\int_t^{\infty} f(\tau,(x_{n-1})_\tau)d\tau + \int_{t_1}^{t} p(\tau)f(\tau,(x_{n-1})_\tau)d\tau \quad t_1 \leq t \leq \infty \tag{10}$$

$$x_n(t) = x_n(t_1) \qquad t_1 - h \leq + < t_1\,,$$

where for the superlinear case, we choose t_1 S.t. $\int_{t_1}^{\infty} f(\tau, cp(\tau+s))|d\tau \leq \frac{|c|}{2}$ and for the sublinear case, we choose t_1 S.t. $\int_{t_1}^{\infty} |f(\tau, \frac{c}{2}p(\tau+s))|d\tau \leq |c|/2$.

Usind assumptions (i) - (iv) or (iv)', one can show by induction

$$|c|/2\; p(t) \leq x_n(t)\ \text{sign}\ c \leq |c|p(t) \quad ; \quad t \geq t_1 - h \tag{11}$$

$n = 0,1,2,\ldots.$

and $\quad |x_n(t)| \leq |x_{n+1}(t)|$ or $|x_{n+1}(t)| \leq x_n(t)|$, $t \geq t_1 - h$, $n = 0,1,2,\ldots.$

It follows that the sequence $\{x_n(t)\}$ converges to some function $x(t)$. Furthermore

$$|c|/2\; p(t) \leq x(t) \cdot \text{Sign}\ c \leq |c|p(t)\,. \tag{12}$$

We now show $x(t)$ is a solution of the integral equation (9). For any fixed t, we choose $T>t$ such that $\int_T^{\infty}|f(\tau, cp(\tau+s))|d\tau < \frac{\varepsilon}{2p(T)}$.

or

$(\int_T^{\infty}|f(\tau,\frac{c}{2}p(\tau+s))|d\tau < \frac{\varepsilon}{2p(T)})$. Then

$$|x_n(t) - \tfrac{c}{2}p(t) - p(t)\int_t^{\infty} f(\tau,x_\tau)d\tau - \int_{t_1}^{t} p(\tau)f(\tau,x_\tau)d\tau|$$

$$\leq p(T)\int_{t_1}^{T} |f(\tau,x_\tau) - f(\tau,(x_{n-1})_\tau)|d\tau + p(T)\int_T^{\infty}|f(\tau,(x)_\tau) - f(\tau,(x_{n-1})_\tau)|d\tau$$

$$\leq p(t)\int_{t_1}^{T} |f(\tau,x_\tau)-f(\tau,(x_{n-1})_\tau)|d\tau + \varepsilon.$$

Case II: $p(t)$ converges.
We define $\{x_n(t)\}$ as

$$x_o(t) = \begin{cases} \frac{c}{2}p(t) & \text{if (1) is supperlinear; } t \geq t_1 - h \\ cp(t) & \text{if (1) is sublinear; } t \geq t_1 - h , \end{cases}$$

$$x_n(t)=\frac{c}{2}p(t)+\int_t^{\infty}p(\tau)f(\tau,(x_{n-1})_\tau)d\tau+p(t)\int_{t_1}^{t} f(\tau,(x_{n-1})_\tau)d\tau \quad t_1 \leq t < \infty , \tag{13}$$

$$x_n(t) = x_n(t_1) \qquad t_1-h \leq t < t_1 ,$$

where t_1 is chosen so that $\int_{t_1}^{\infty} |f(\tau,cp(\tau+s))|d\tau \leq |c|/2$ or $\int_{t_1}^{\infty} |f(\tau,\frac{c}{2}p(\tau+s))|d\tau \leq \frac{|c|}{2}$. for (1) be superlinear or sublinear just as above, we show that equation (13) has a solution $x(t)$ satisfying

$$\frac{|c|}{2}\, p(t) \leq x(t) \text{ sign } c \leq |c|p(t) ,$$

i.e. $x(t)$ is a nonoscillatory function. The theorem is proved.

Theorem 2. If in addition to the conditions (i) - (iv) or (iv)', we assume

$$\int_{t_o}^{\infty} |f(\tau,c)|d\tau = \infty \tag{14}$$

for every non-zero constant c, then

a) for superlinear case, any solution of (1) is either oscillatory or tends to zero as $t \to \infty$.

b) for sublinear case any bounded solution of (1) is oscillatory.

Proof. Assume (1) is superlinear, then if $x(t)$ is a solution which is not oscillatory, then for $t \geq t_1$ (t_1 sufficiently large) it preserves its sign. So let us assume it is positive. Therefore $x'(t)$ is either positive or negative.

I) If $x'(t) > 0$, we have $|x(t)| \geq |x(t_1)|$ which implies

$$r(t)|x'(t)| = r(t_1)|x'(t_1)| - \int_{t_1}^{t} |f(\tau,x_\tau)|d\tau.$$

Now, since if $|x(t)| > |x(t_1)|$ it is also $|x_t| > |x(t_1)|$. We have

$$r(t)|x'(t)| \leq r(t_1)|x'(t_1)| - \int_{t_1}^{t} |f(\tau,x(t_1))|d\tau \ ;$$

letting $t \to \infty$ we get a contradiction.

II) If $x'(t) < 0$. Then $p(t)$ must converge. Otherwise

$$r(t)\ x'(t) \leq r(t_1)\ x'(t_1) \text{ or } r(t)x'(t) \leq r(t_1)x'(t_1) \cdot \frac{1}{r(t)},$$

which implies

$$x(t)-x(T) \leq r(t_1)x'(t_1)\int_T^t \frac{d\tau}{r(\tau)} \text{ or } r(t_1)|x'(t_1)|\int_T^t \frac{d\tau}{r(\tau)} \leq x(T).$$

Letting $t \to \infty$, we obtain a contradiction.

Now, since $x(t) > 0$ and $x'(t) < 0$, it follows either $\lim_{t\to\infty} x(t) = 0$ or $\lim_{t\to\infty} x(t) = M$ (a constant). In the first case we have proved the theorem. For the second case, there are constants c_1 and c_2 such that for $t \geq t_1$, $c_1 \leq x(t) \leq c_2$ which implies $c_1 \leq x_t \leq c_2$. Therefore

$$x'(t) \leq -\frac{1}{r(t)} \int_{t_1}^{t} |f(\tau,c)|d\tau$$

or

$$x(t) - x(T) \leq -\int_T^t \frac{1}{r(s)} \{\int_{t_1}^{s} |f(\tau,c)|d\tau\}ds.$$

Now by mean value theorem of integral, we obtain

$$\int_r^t \frac{ds}{r(s)} \cdot \int_{t_1}^{s^*} |f(\tau,c)|d\tau \leq x(T)$$

$$T < s^* < t \qquad ;$$

letting $t \to \infty$, we obtain a contradiction. Hence the theorem is proven.

(II) If (1) is sublinear, with a calculation just as above we show $x(t)$ is oscillatory.

Theorem 3. If for any nonzero constant c, we can find constants $\lambda \neq 0$ and $M > 0$ depending on c, such that the inequality

$$|f(t,c)| \geq M\,|f(t,\lambda p(t))| \tag{15}$$

is satisfied for t sufficiently large, then

a) for (1) superlinear and p(t) diverges, condition (14) is necessary and sufficient in order (1) to be oscillatory.

b) for (1) sublinear and p(t) converges condition (14) is again necessarry and sufficient in order (1) to be oscillatory.

Proof. The sufficiency of the condition follows from theorem (2). We prove the necessity of the condition. For this we show that if $\int_{t_0}^{\infty} |f(\tau,c)|\,d\tau < \infty$, then (1) has at least one nonoscillatory solution. Indeed, according to condition (15) we have

$$\int_{t_0}^{t} |f(\tau,\lambda p(\tau))|\,d\tau \leq \frac{1}{M}\int_{t_0}^{t} |f(\tau,c)|\,d\tau < \infty.$$

But then by theorem 1. equation (1) has at least one nonoscillatory solution. This proves the theorem.

Theorem 4. Assume (1) is superlinear and p(t) converges, then (4) is a necessary and sufficient condition in order (1) to be oscillatory.

Proof. The necessity of the condition follows from theorem 1. We prove the sufficiency. Assume $x(t)$ is a solution of (1) which is not oscillatory. Therefore, we can assume $x(t) > 0$ for $t \geq t_1$. Therefore $x'(t)$ is either positive or negative for $t \geq t_1$.

I) Assume $x'(t) > 0$ for $t \geq t_1$. This implies $|x(t)| \geq |x(t_1)|$ and there exists some constant $c \neq 0$ such that $|x(t)| \geq |x(t_1)| \geq cp(t)$; this implies

$$r(t)|x'(t)| = r(t_0)|x'(t_0)| - \int_{t_0}^{t} |f(\tau,x_\tau)|\,d\tau.$$

or

$$r(t)|x'(t)| < f(t_0)|x'(t_0)| - \int_{t_0}^{t} |f(\tau,cp(\tau+s))|\,d\tau.$$

II) Assume $x'(t) < o$ for $t \geq t_1$. On one hand

$$r(t)x'(t) \leq r(t_1)x'(t_1)$$

or

$$x(t)-x(T) \leq r(t_1)x'(t_1) \int_T^t \frac{d\tau}{r(\tau)}$$

which implies $x(T) \geq c\, p(T)$ $(c = r(t_1)x'(t_1))$. Now since $T > t_1$ is any real number, we have $x(t) \geq c\, p(t)$ for $t > t_1$.
On the other hand

$$x'(t) \leq -\frac{1}{r(t)} \int_{t_1}^{t} |f(\tau,cp(\tau+s))|d\tau$$

or

$$x(t) - x(T) \leq -\int_T^t \frac{1}{r(u)} \{\int_{t_1}^{u} |f(\tau,ep(\tau+s))|d\tau\}du$$

which by mean value theorem of integral leads to

$$\{\int_T^t \frac{du}{r(u)}\} \cdot \int_{t_1}^{u^*} |f(\tau,ep(\tau+s))|d\tau \leq x(T)$$

$$T < u^* < t ;$$

letting $t \to \infty$, we obtain a contradiction. Hence the theorem is proven.

Theorem 5. If instead of condition (iv) or (iv)', we assume

$$\text{(v)} \quad \frac{|f(t,y_1)|}{|y_1|} \leq \frac{|f(t,y_2)|}{|y_2|} \quad \text{when} \quad |y_1| \leq |y_2|$$

for each $t \in |t_o - h,\infty)$. then a sufficient condition in order (1) to be oscillatory is that

$$\limsup_{t\to\infty} \int_{t_o}^{t} p(\tau)\{ \frac{f(\tau,c)}{c} - \frac{1}{4r(\tau)p^2(\tau)}\} d\tau = \infty \tag{16}$$

for any nonzero constant c, provided that p(t) is divergent.

Proof. Let x(t) be a nonoscillatory solution of (1). Following the first part of the argument in the previous theorems, we choose t_1 and T as before. Next we multiply both sides of Equation (1) by $\frac{p(t)}{x(t)}$ and integrate from T to t :

$$p(t)r(t)\frac{x'(t)}{x(t)}\int_T^t\frac{x'(\tau)}{x(\tau)}d\tau+\int_T^t r\cdot p\left(\frac{x'(\tau)}{x(\tau)}\right)^2 d\tau+\int_T^t\frac{p}{x(\tau)}f(\tau,x_\tau)d\tau=p(Tr(T)\frac{x'(T)}{x(T)} .$$

Applying the inequality

$$\frac{x'(t)}{x(t)} \leq r\ p\ \left(\frac{x'(t)}{x(t)}\right)^2 + \frac{1}{4rp} ,$$

we obtain

$$\int_T^t p(\tau)\ \left\{\frac{f(\tau,x_\tau)}{x(\tau)} - \frac{1}{4rp^2}\right\}\ d\tau \leq p(T)r(T)\ \frac{x'(T)}{x(T)} .$$

Now, using the fact $|x(t)| \geq |x(T)|$, $|x_t| \geq |x(T)|$ and condition (v), we get

$$\int_T^t p(\tau)\ \left\{\frac{f(\tau,x(T))}{x(T)} - \frac{1}{4rp^2}\right\}\ d\tau \leq p(T)r(T)\ \frac{x'(T)}{x(T)},$$

which contradicts (16). This completes the proof of theorem 5.

REFERENCES

(1) IZYUMOVA, D. V.: On the Conditions of Oscillation and Non-oscillation of Solutions of Nonlinear Second Order Differential Equations 'Differentsial'nye uravneniya', 2(1966), 1572-1586.

(2) ATKINSON, F. E.: On Second-order Nonlinear Oscillations, Pacific J. Math., 5 (1955), 643-647.

(3) EL,SOGOL,TS, L. E.: Introduction to the Theory of Differential Equations with Deviating Arguments, Halden-Day San Francisco, 1966.

(4) Ya. V. BYKOV - L. Ya. BYKOVA, E. I. SEVCOV : Sufficient Conditions for Oscillation of Solutions of Nonlinear Differential Equations with Deviating Argument, 'Differencial'nye Uravneniya', 9, (1973), pp. 1555-1560 (Russian).

(5) B. MEHRI : On the Conditions for the Oscillation of Solutions of Non-linear Third order Differential Equations to appear in "CASOPIS PRO PESTOVANI MATEMATIKY", CZECHOSLOVAK MATHEMATICAL JOURNAL.

Eine spezielle Integralgleichung erster Art

Theodor Meis

Mathematisches Institut der Universität zu Köln,

Weyertal 86-90, D-5000 Köln 41

Bei den Auswertungen von Experimenten der Atomphysik stößt man in gewissen Fällen auf einen speziellen Typ von Volterraschen Integralgleichungen erster Art (vgl. U. Freitag [5] und R. Esser [1]). Die gesuchte Funktion y kommt unter dem Integral mit dem Argument t und einem zweiten Argument a(x,t) vor. Es handelt sich darum eigentlich um Integrofunktionalgleichungen.

Aufgabenstellung IF1

Gesucht ist eine Funktion $y:[0,1]\to\mathbb{R}$ mit

$$g(x) = \int_0^x k(x,t,y(t),y(a(x,t)))dt \ , \quad x \in [0,1]$$

$$g \in C^2([0,1],\mathbb{R}), \quad k \in C^2([0,1]^2\times\mathbb{R}^2, \mathbb{R})$$

$$a \in C^2([0,1]^2,[0,1])$$

$$0 \leq a(x,t) \leq \max(rx,x-s) \ , \quad x \in [0,1], t \in [0,x]$$

$$r \in (0,1) \text{ und } s \in \mathbb{R}_+ \text{ fest}$$

$$a_t(x,t) \geq \alpha > 0 \ , \quad x \in [0,1], t \in [0,x]$$

$$k_u(x,x,u,v) \geq \beta > 0 \ , \quad x \in [0,1], u,v \in \mathbb{R}$$

$$\alpha,\beta \in \mathbb{R}_+ \text{ fest.}$$

($C^l(A,B)$ = Menge der l-mal stetig differenzierbaren Funktionen $A\to B$, $C(A,B) = C^0(A,B)$.) Wegen der Bedingung $a(x,t) \geq 0$ geht im Gegensatz zu vielen ähnlichen Problemen hier keine Vorgeschichte von y (y(x) mit $x < 0$)

ein. Das Argument $a(x,t)$ ist gegenüber x verzögert

$$x-a(x,t) \geq \min((1-r)x,s).$$

Wenn das Argument $y(a(x,t))$ in k fehlt, kann die Gleichung durch Differentiation sofort in eine Integralgleichung zweiter Art überführt werden.

Im Spezialfall

$$k(x,t,u,v) = k_1(x,t,u) + k_2(x,t,v)$$

kann man das Integral in zwei Integrale aufspalten. Eine Substitution $\hat{t} = a(x,t)$ im zweiten Summand und anschließende Differentiation der ganzen Gleichung führt auf eine Gleichung vom weiter unten behandelten Typ IF2.

Im allgemeinen Fall liegt es nahe, k für y in der Nähe einer Funktion $w \in C^2([0,1],\mathbb{R})$ zu linearisieren und dann das Integral wieder in zwei Summanden aufzuspalten. Tatsächlich ist R. Esser in einem konkreten Anwendungsfall so vorgegangen. Mit Hilfe der von ihm entwickelten numerischen Verfahren zur Lösung der auftretenden Integralgleichungen zweiter Art kam er zu befriedigenden numerischen Resultaten.

Es blieb aber ungeklärt, in welchen Räumen und unter welchen Bedingungen Existenz und Eindeutigkeit der Lösung von IF1 gewährleistet werden können.

Hier möchte ich zunächst zeigen, daß IF1 auf ein System von Integrofunktionalgleichungen vom Typ IF2 umgeschrieben werden kann. Anschließend folgen zwei Existenz- und Eindeutigkeitssätze für IF2 und am Ende ein kurzer Bericht über die von R. Esser angewandten numerischen Methoden.

In der folgenden Rechnung benutze ich die Abkürzungen

$$(Axt) = (x,t,y(t),y(a(x,t)))$$

$$(Axx) = (x,x,y(x),y(a(x,x)))$$

$$(Ax0) = (x,0,y(0),y(a(x,0))).$$

Bei der Bildung von partiellen Ableitungen wird k als $k(x,t,u,v)$ aufgefaßt.

Aus der Gleichung

$$g(x) = \int_0^x k(Axt)dt$$

erhält man durch Differentiation

$$g'(x) = k(Axx) + \int_0^x k_x(Axt)dt + \int_0^x k_v(Axt)y'(a(x,t))a_x(x,t)dt.$$

Wir setzen nun weiter voraus, daß $y(0) = \gamma$ eindeutig aus der Gleichung $g'(0) = k(0,0,\gamma,\gamma)$ bestimmt werden kann. Durch partielle Integration formen wir dann die Integralgleichung noch etwas um. Dazu setzen wir $b(x,t) = a_x(x,t)/a_t(x,t)$ und erhalten

$$k_v(Axt)y'(a(x,t))a_x(x,t) = \frac{d}{dt}[k(Axt)]b(x,t)$$
$$-k_t(Axt)b(x,t)-k_u(Axt)y'(t)b(x,t)$$

und die Integralgleichung

$$g'(x) = k(Axx)(1+b(x,x))-k(Ax0)b(x,0)$$
$$+\int_0^x [k_x(Axt)-k(Axt)b_t(x,t)-k_t(Axt)b(x,t)$$
$$-k_u(Axt)b(x,t)y'(t)]dt.$$

Sie hat den Vorteil, daß bei nochmaliger Differentiation kein Term entsteht, der y" enthält. Für $y \in C^1([0,1],\mathbb{R})$ ist also eine nochmalige

Differentiation der Gleichung möglich. Wir verzichten darauf, die differenzierte Gleichung ganz hinzuschreiben und notieren nur die Summanden außerhalb des Integrals, die y' enthalten:

$$k_u(Axx)y'(x)$$
$$+k_v(Axx)(1+b(x,x))y'(a(x,x))(a_x(x,x)+a_t(x,x))$$
$$-k_v(Ax0)b(x,0)y'(a(x,0))a_x(x,0)$$

Wegen $k_u(Axx) \geq \beta > 0$ ist die Gleichung also nach $y'(x)$ auflösbar.

Hinzu kommt die Gleichung

$$y(x) = \gamma + \int_0^x y'(t)dt.$$

Beide Gleichungen zusammen bilden ein System von Integrofunktionalgleichungen zweiter Art für y und y'.

Wir wollen nun Systeme dieser Art hinsichtlich der Existenz und Eindeutigkeit der Lösungen untersuchen.

Aufgabenstellung IF2

Gesucht ist eine Funktion $y \in C([0,1],\mathbb{R}^m)$, die einer Integrofunktionalgleichung zweiter Art genügt:

$$y(x) = F(x,By(x),Ky(x))$$

Bezeichnungen:

$$By(x) = (y(b_1(x)),\ldots,y(b_n(x)))^T$$
$$Ky(x) = \int_0^x k(x,t,y(t),y(a(x,t)))dt$$
$$F \in C([0,1]\times\mathbb{R}^{m(n+1)},\mathbb{R}^m), \quad k \in C([0,1]^2\times\mathbb{R}^{2m},\mathbb{R}^m)$$
$$a \in C([0,1]^2,[0,1])$$
$$b_\nu \in C([0,1],[0,1]) , \quad \nu = 1(1)n$$

In Anlehnung an die ursprüngliche Aufgabe wird für die verzögerten

Argumente a und b_ν im folgenden generell vorausgesetzt:

$$0 \leq a(x,t) \leq \max(rx, x-s) \quad , \quad x \in [0,1], t \in [0,x]$$

$$0 \leq b_\nu(x) \leq \max(rx, x-s) \quad , \quad \nu = 1(1)n, x \in [0,1]$$

$r \in (0,1)$ und $s \in \mathbb{R}_+$ fest. □

Die Voraussetzungen reichen offensichtlich nicht, um die eindeutige Lösbarkeit von IF2 auch nur lokal sicherzustellen. Notwendig sind vielmehr noch Lipschitzbedingungen an F und k. Sie können in verschiedener Weise formuliert werden. Wir wollen die Bedingungen in zwei Gruppen zusammenfassen. Die erste Gruppe bezieht sich zwar auf die nichtverzögerten und die verzögerten Argumente von F und k, hat aber streng lokalen Charakter. ($\|\cdots\|$ = Maximumnorm)

Bedingungen LB1: Es gibt $c \in \mathbb{R}^m$ und $\varepsilon > 0$ mit folgenden Eigenschaften:

i) $c = F(0,z,o)$ mit $z = (c,\ldots,c)^T \in \mathbb{R}^{mn}$

und $o = (0,\ldots,0)^T \in \mathbb{R}^m$

ii) Für alle $v,\tilde{v} \in \mathbb{R}^{mn}$, $w,\tilde{w} \in \mathbb{R}^m$ und $x \in [0,1]$ mit

$\|v-z\| < \varepsilon$, $\|\tilde{v}-z\| < \varepsilon$, $\|w\| < \varepsilon$, $\|\tilde{w}\| < \varepsilon$, $x \in [0,\varepsilon]$

gilt für festes $L_1 \in (0,1)$ und festes $L_2 \in \mathbb{R}_+$

$\|F(x,v,w)-F(x,\tilde{v},\tilde{w})\| < L_1 \|v-\tilde{v}\| + L_2 \|w-\tilde{w}\|$.

iii) Für alle $u,\tilde{u},w,\tilde{w} \in \mathbb{R}^m$, $x \in [0,1]$ und $t \in [0,x]$ mit

$\|u-c\| < \varepsilon$, $\|\tilde{u}-c\| < \varepsilon$, $\|w-c\| < \varepsilon$, $\|\tilde{w}-c\| < \varepsilon$, $x \in [0,\varepsilon]$

gilt für festes L_3 und $L_4 \in \mathbb{R}_+$

$\|k(x,t,u,w) - k(x,t,\tilde{u},\tilde{w})\| \leq L_3 \|u-\tilde{u}\| + L_4 \|w-\tilde{w}\|$. □

Besondere Beachtung verdient die Forderung $L_1 < 1$. Sie besagt, daß F lokal nicht zu stark von den Termen außerhalb des Integrals abhängen darf. Im übrigen sind die lokalen Lipschitzbedingungen ii) und iii) schon erfüllt, wenn F im Punkte (0,z,o) differenzierbar ist.

Es zeigt sich, daß man globale Lipschitzbedingungen an F und k nur hinsichtlich der nichtverzögerten Argumente braucht. Die Liptschitzkonstanten in diesen Ungleichungen dürfen von den verzögerten Argumenten abhängen.

<u>Bedingungen LB2:</u>

i) Für alle $x \in [0,1]$, $v \in \mathbb{R}^{mn}$ und $w,\tilde{w} \in \mathbb{R}^m$ gilt mit festem $h_1 \in C(\mathbb{R}^{mn}, \mathbb{R}_+)$: $\|F(x,v,w)-F(x,v,\tilde{w})\| < h_1(v)\|w-\tilde{w}\|$

ii) Für alle $x \in [0,1]$, $t \in [0,x]$, $u,\tilde{u} \in \mathbb{R}^m$, $w \in \mathbb{R}^m$ gilt mit festem $h_2 \in C(\mathbb{R}^m, \mathbb{R}_+)$: $\|k(x,t,u,w)-k(x,t,\tilde{u},w)\| \leq h_2(w)\|u-\tilde{u}\|$. □

LB2 impliziert nicht LB1.

<u>Satz 1:</u> F und k genügen den Bedingungen LB1. Dann gibt es $\eta \in (0,1]$, so daß IF2 genau eine Lösung $y \in C([0,\eta],\mathbb{R}^m)$ mit $y(0) = c$ besitzt. □

<u>Satz 2:</u> F und k genügen den Bedingungen LB1 und LB2. Dann hat IF2 genau eine Lösung in $C([0,1],\mathbb{R}^m)$ mit $y(0) = c$. □

Bei beiden Sätzen handelt es sich um Verallgemeinerungen eines Satzes aus der Dissertation von R. Esser [1]. Wie dort, benutzen wir beim Beweis den Banachschen Fixpunktsatz und eine gewichtete Maximumnorm in den Räumen $C([0,1],\mathbb{R}^m)$ bzw. $C([0,\delta],\mathbb{R}^m)$. Für $\alpha \geq 0$ definieren wir

$$\|y\|_\alpha = \max_{x\in[0,\delta]} \|e^{-\alpha(x^2+1)} y(x)\| .$$

Alle diese Normen sind äquivalent und erzeugen die gleiche Topologie in $C([0,\delta],\mathbb{R}^m)$.

<u>Hilfssatz:</u> Für alle $x \in [0,\delta], t \in [0,x], u,\tilde{u} \in \mathbb{R}^m$ und $w,\tilde{w} \in \mathbb{R}^m$ gelte

$$\| k(x,t,u,w)-k(x,t,\tilde{u},\tilde{w})\| \leq L_5 \|u-\tilde{u}\| + L_6 \|w-\tilde{w}\|$$

mit festen Konstanten L_5 und $L_6 \in \mathbb{R}_+$. Dann gibt es zu jedem $\beta > 0$ ein $\alpha \geq 0$, so daß für alle $y,\tilde{y} \in C([0,\delta],\mathbb{R}^m)$ gilt

$$\|Ky-K\tilde{y}\|_\alpha \leq \beta \|y-y\|_\alpha . \quad \square$$

Den Beweis des Hilfssatzes findet man bei R. Esser.

Als Vorbereitung für die Beweise der Sätze 1 und 2 definieren wir noch für jedes $\delta > 0$ eine Abbildung $p_\delta \in C(\mathbb{R}^1, \mathbb{R}^1)$. Sei

$$u = (u_1,\ldots,u_1)^T \in \mathbb{R}^1$$

dann ist

$$p_\delta(u) = (\hat{u}_1,\ldots,\hat{u}_1) \in \mathbb{R}^1$$

$$\hat{u}_\nu = \begin{cases} u_\nu \text{ für } |u_\nu| \leq \delta \\ \delta \operatorname{sign}(u_\nu) \text{ sonst.} \end{cases}$$

Man kann leicht zeigen

α) p_δ ist stetig

β) $p_\delta(u) = u$ für alle u mit $\|u\| \leq \delta$

γ) $\|p_\delta(u)\| \leq \delta$ für alle u

δ) $\|p_\delta(u)-p_\delta(\tilde{u})\| \leq \|u-\tilde{u}\|$ für alle $u,\tilde{u} \in \mathbb{R}^1$

<u>Beweis von Satz 1</u>: Mit den c,z und ε aus LB1 definieren wir

$$\hat{F}(x,v,w) = F(x,z+p_\varepsilon(v-z),p_\varepsilon(w))$$

$$\hat{k}(x,t,u,w) = k(x,t,c+p_\varepsilon(u-c),c+p_\varepsilon(w-c))$$

$$\hat{K}y(x) = \int_0^x \hat{k}(x,t,y(t),y(a(x,t)))dt.$$

$\hat{F},\hat{k}$ sind stetig; in einer Umgebung der Punkte $(0,z,o)$ bzw. $(0,0,c,c)$ stimmen F und $\hat{F}$ bzw. k und $\hat{k}$ überein. Darum ist jede stetige Lösung von

$$y(x) = F(x,By(x),Ky(x))$$

mit $y(0) = c$ in einem hinreichend kleinen Intervall $[0,\eta]$ auch Lösung von

$$y(x) = \hat{F}(x,By(x),\hat{K}y(x))$$

und umgekehrt. Es genügt demnach zu beweisen: Sei $\delta \in (0,\varepsilon]$. Dann hat die Gleichung

$$y(x) = \hat{F}(x,By(x),\hat{K}y(x)) \ , \quad x \in [0,\delta]$$

genau eine Lösung $y \in C([0,\delta],\mathbb{R}^m)$ mit $y(0) = c$.

Die Abbildungen $\hat{F}$ und $\hat{k}$ sind so konstruiert, daß sie folgenden globalen Lipschitzbedingungen genügen

α) $\|\hat{F}(x,v,w)-\hat{F}(x,\tilde{v},\tilde{w})\| \leq L_1 \|v-\tilde{v}\| + L_2 \|w-\tilde{w}\|$

für alle $x \in [0,\varepsilon], v,\tilde{v} \in \mathbb{R}^{mn}$ und $w,\tilde{w} \in \mathbb{R}^m$

β) $\|\hat{k}(x,t,u,w)-\hat{k}(x,t,\tilde{u},\tilde{w})\| \leq L_3 \|u-\tilde{u}\| + L_4 \|w-\tilde{w}\|$

für alle $x \in [0,\varepsilon], t \in [0,x], u,\tilde{u},w,\tilde{w} \in \mathbb{R}^m$.

Die Konstanten in den Ungleichungen sind die gleichen wie in LB1, insbesondere gilt $L_1 < 1$. Wir wählen nun $\alpha \geq 0$ so, daß gemäß dem zitierten Hilfssatz gilt

$$\|\hat{K}y-\hat{K}\tilde{y}\|_\alpha \leq \beta \|y-\tilde{y}\|_\alpha$$

und $\qquad L = L_1 + L_2\, \beta < 1.$

Es folgt für alle $y,\tilde{y} \in C([0,\delta],\mathbb{R}^m)$

$$\|\hat{F}(x,By(x),\hat{K}y(x))-\hat{F}(x,B\tilde{y}(x),\hat{K}\tilde{y}(x))\| \leq L_1 \|By(x)-B\tilde{y}(x)\| + L_2 \|\hat{K}y(x)-\hat{K}\tilde{y}(x)\|$$

$$\|\hat{F}(\cdot,By(\cdot),\hat{K}y(\cdot))-\hat{F}(\cdot,B\tilde{y}(\cdot),\hat{K}\tilde{y}(\cdot))\|_\alpha \leq L_1 \|By-B\tilde{y}\|_\alpha + L_2 \|Ky-K\tilde{y}\|_\alpha .$$

Wegen $b_\nu(x) \leq x$, $\nu = 1(1)n$ ergibt sich

$$\|By-B\tilde{y}\|_\alpha \leq \|y-\tilde{y}\|_\alpha$$

$$\|\hat{F}(\cdot,By(\cdot),\hat{K}y(\cdot))-\hat{F}(\cdot,By(\cdot),\hat{K}y(\cdot))\|_\alpha \leq L\|y-\tilde{y}\|_\alpha .$$

Die Abbildung $y \to \hat{F}(\cdot,B\tilde{y}(\cdot),\hat{K}\tilde{y}(\cdot))$ ist demnach in der Norm $\|\cdots\|_\alpha$ kontrahierend und hat nach dem Banachschen Fixpunktsatz genau einen Fixpunkt in $C([0,\delta],\mathbb{R}^m)$. Wir nennen ihn wieder y. Bleibt zu beweisen $y(0) = c$. Es gilt aber nach LB1

$$\|y(0)-c\| = \|\hat{F}(0,By(0),o)-\hat{F}(0,z,o)\| \leq L_1 \|By(0)-z\| = L_1\|y(0)-c\| .$$

Wegen $L_1 < 1$ bedeutet das: $y(0) = c$. □

<u>Beweis von Satz 2:</u> Nach Satz 1 ist die Aufgabe in $[0,\eta]$ eindeutig lösbar. Diese Teillösung sei e. Für $x \in [0,\eta+\delta]$ mit $\delta = \min((1-r)\eta,s)$ liegen die verzögerten Argumente $b_\nu(x),\nu = 1(1)n$ und $a(x,t)$ im Intervall $[0,\eta]$. Wir lösen darum zunächst die Aufgabe

$$y(x) = F(x,T(x),Sy(x))$$

$$y(0) = c$$

$$T(x) = (e(b_1(x)),\ldots,e(b_n(x)))^T$$

$$Sy(x) = \int_0^x k(x,t,y(t),e(a(x,t)))dt.$$

$e(b_\nu(x)),\nu = 1(1)n$ und $e(a(x,t))$ sind bekannte feste Funktionen. Es sei

$$M = \sup_{x\in[0,\eta]} \|e(x)\|$$

$$N_1 = \sup_{\|v\| \leq M} h_1(v)$$

$$N_2 = \sup_{\|w\| \leq M} h_2(w).$$

Aus LB2 folgt

$$\|F(x,T(x),w)-F(x,T(x),\tilde{w})\| < N_1 \|w-\tilde{w}\|$$

$$\| k(x,t,u,e(a(x,t)))-k(x,t,\tilde{u},e(a(x,t)))\| \le N_2\|u-\tilde{u}\|.$$

$\tilde{k}(x,t,u) = k(x,t,u,e(a(x,t)))$ erfüllt die Voraussetzungen des Hilfssatzes mit $L_5 = N_2$ und L_6 beliebig. Es gibt darum ein $\alpha \geq 0$ mit

$$\|Sy-S\tilde{y}\|_\alpha \leq (2N_1)^{-1}\|y-\tilde{y}\|_\alpha$$

für alle $y,\tilde{y} \in C([0,\eta+\delta],\mathbb{R}^m)$. In der Norm $\|\cdots\|_\alpha$ ist die Abbildung $y \to F(x,T(x),Sy(x))$ kontrahierend. Es gibt darum genau eine Lösung e_1 der Gleichung

$$y(x) = F(x,T(x),Sy(x))$$

im Intervall $[0,\eta+\delta]$. Die Beschränkung der Lösung auf $[0,\eta]$ ist e. Also ist e_1 auch Lösung von

$$y(x) = F(x,By(x),Ky(x))$$
$$y(0) = c$$

im Intervall $[0,\eta+\delta]$. Sei $\tilde{e}_1$ eine andere Lösung dieser Aufgabe. Die Beschränkung von e_1 auf $[0,\eta]$ ist nach Satz 1 e. Dann ist $\tilde{e}_1$ aber Lösung von $y(x) = F(x,T(x),Sy(x))$, also $\tilde{e}_1 = e_1$. Da für $x \in [0,\eta+2\delta]$ die verzögerten Argumente in $[0,\eta+\delta]$ liegen, zeigt man analog, daß IF2 im Intervall $[0,\eta+2\delta]$ eindeutig lösbar ist. Nach endlich vielen Schritten erhält man: IF2 ist eindeutig lösbar in $[0,1]$. □

Anhand einiger Beispiele mit $m = n = 1$ möchte ich noch die Bedeutung der Forderung $L_1 < 1$ aus LB1 erläutern.

Bei der Funktionalgleichung

$$y(x) = y(x/2)$$

kann man $L_1 = 1$ wählen. Jede Konstante ist Lösung.

Auch bei der Gleichung

$$y(x) = -y(x/2)$$

ist LB1 nicht erfüllt. Diese Gleichung hat aber genau eine stetig differenzierbare Lösung in [0,1]. Wenn y nämlich differenzierbar ist, ist die Gleichung äquivalent zu dem System

$$y'(x) = -\frac{1}{2}\, y'(x/2)$$
$$y(x) = \int_0^x y'(t)dt\ .$$

Für dieses System gelten LB1 und LB2; Satz 2 ergibt die eindeutige Lösbarkeit ($y(x) = 0$).

Die Gleichung

$$y(x) = 3y(x/2)$$

ist auf das System

$$y''(x) = \frac{3}{4}\, y''(x/2)$$
$$y'(x) = \int_0^x y''(t)dt$$
$$y(x) = \int_0^x y'(t)dt$$

zurückzuführen. Nach Satz 2 gibt es darum genau eine zweimal stetig differenzierbare Lösung ($y(x) = 0$).

Die Gleichung

$$y(x) = 2y(x/2)$$

hat aber jede Funktion $y(x) = \gamma x$ mit $\gamma \in \mathbb{R}$ zu Lösung.

Die angegebenen Lösungen des ersten bzw. letzten Beispiels sind nach Satz 2 die einzigen stetig differenzierbaren bzw. zweimal stetig differenzierbaren Lösungen dieser Funktionalgleichungen.

Das letzte Beispiel

$$y(x) = 2y(x/2)$$

hat aber noch andere stetige Lösungen, z.B.

$$y(x) = \begin{cases} 2\cdot 2^{-n} & \text{für } \frac{3}{2}\cdot 2^{-n} \le x \le 2\cdot 2^{-n} \\ 2^{-n}+2(x-2^{-n}) & \text{für } 2^{-n} \le x \le \frac{3}{2}\cdot 2^{-n} \qquad n = 1(1)\infty \\ 0 & \text{für } x = 0 \end{cases}$$

Numerische Verfahren

Die numerischen Verfahren zur Lösung Volterrascher Integralgleichungen haben im allgemeinen eine große Ähnlichkeit mit den Verfahren zur Lösung von Anfangswertaufgaben für gewöhnliche Differentialgleichungen (vgl. etwa F. De Hoog und R. Weiss [2], M.E.A. El Tom [4] sowie den Sammelband L.M. Delves und J. Walsh [3]). Beim System IF2 liegen wegen der verzögerten Argumente $b_\nu(x)$ und $a(x,t)$ besondere Verhältnisse vor. Wenn t und x eine Menge diskreter äquidistanter Punkte durchlaufen, liegen $b_\nu(x)$ und $a(x,t)$ doch zwischen diesen Stützpunkten. Die Lösung der Gleichung erfordert darum die Anwendung eines geeigneten Interpolationsverfahren.

R. Esser benutzte stückweise Hermite-Interpolation. Seien $j = 0$ oder $j \in \mathbb{N}$, $N \in \mathbb{N}$ und $h = 1/N$ fest vorgegeben. y wird dann approximiert durch eine Funktion $f \in C^j([0,1],\mathbb{R}^m)$, deren Beschränkung auf jedes Teilintervall $[(\mu-1)h,\mu h]$, $\mu = 1(1)N$ ein Polynom vom Grade $< 2j + 2$ ist. Alle f dieser Art bilden einen Vektorraum der Dimension $(N+1)(j+1)m$. Es gibt eine Basis dieses Vektorraums mit Funktionen, deren Träger Intervalle der Länge 2h oder h sind. Das ist ein wesentlicher Vorteil gegenüber kubischen Splines. Die Werte $f^{(\nu)}(0)$, $\nu = 0(1)j$ werden direkt aus der Integralgleichung und ihren Ableitungen berechnet. Also bleiben $N(j+1)m$ Koeffizienten zu bestimmen.

Der Operator K wird folgendermaßen durch einen Operator Q approximiert: Man teilt das Integral in Teilintegrale über die Intervalle $[(\mu-1)h,\mu h]$ auf. Das letzte Teilintervall kann kürzer sein. In jedem Teilintervall wird das Integral durch die Gaußsche Quadraturformel mit j+1 Stützstellen ersetzt.

Die freien Koeffizienten erhält man nun aus dem Gleichungssystem

$$f(x_{\mu\nu}) = F(x_{\mu\nu},Bf(x_{\mu\nu}),Qf(x_{\mu\nu}))$$

$$x_{\mu\nu} = (\mu-1)h+\eta_\nu h$$

$$\mu = 1(1)N, \qquad \nu = 0(1)j.$$

Die η_ν sind feste Konstanten aus $(1/2,1]$. Praktisch wählt man immer $\eta_j = 1$. Das Gleichungssystem kann schrittweise gelöst werden. In jedem Teilschritt muß man ein Gleichungssystem mit $(j+1)m$ Unbekannten lösen.

Für den Fall

$$F(x,By(x),Ky(x)) = g(x)+\int_0^x k(x,t,y(t),y(a(x,t)))dt$$

und hinreichend große, verschiedene η_ν hat R. Esser die Stabilität der Verfahren bewiesen. Für $j = 0$ und $\eta_0 = 1$ handelt es sich sogar um ein A-stabiles Verfahren. Die Konvergenzordnung ist $2j + 2$. Es gibt also Verfahren dieser Art beliebig hoher Ordnung. Für große j liegen die η_ν allerdings so dicht, daß die Verfahren sehr rundungsfehleranfällig sind. Mit folgenden Verfahren liegen aber sehr positive Erfahrungen vor:

$j=0$	$\eta_o=1$			Ordnung 2
$j=1$	$\eta_o=0.6$	$\eta_1=1$		Ordnung 4
$j=2$	$\eta_o=0.75$	$\eta_1=0.87$	$\eta_2=1$	Ordnung 6

Literatur

[1] Esser, R.: Numerische Lösung einer verallgemeinerten Volterra'schen Integralgleichung zweiter Art. Dissertation, Köln 1976

[2] De Hoog, F. a. Weiss, R.: Implicit Runge-Kutta methods for second kind Volterra integral equations. Numer.Math. 23, 199-213(1975)

[3] Delves, L.M. a. Walsh, J. (Ed.): Numerical solution of integral equations. Oxford: Clarendon 1974

[4] El Tom, M.E.A.: Numerical solution of Volterra integral equations by spline functions. Bit 13, 1-7(1973)

[5] Freitag, U.:Physikalische und mathematische Analyse von Dopplerspektren nach Kernreaktionen. Dissertation, Köln 1974

Ein Zusammenhang zwischen Aufgaben monotoner Art und Intervall-Mathematik

von

Karl Nickel in Freiburg

Meinem verehrten Mathematiklehrer

Professor Wilhelm Schweizer

zu seinem 75ten Geburtstag am 11.11.1976 gewidmet. Ohne seinen Einfluß hätte ich Mathematik nie zu meinem Beruf gewählt.

1. Einleitung

Der Begriff "Aufgaben von monotoner Art" wurde von L.Collatz [5] in die numerische Mathematik eingeführt. Man versteht darunter den folgenden Sachverhalt: Die Menge S sei mit der Ordnungsrelation $\leq$ versehen und bilde damit einen halbgeordneten Raum. Der Operator $T : S \to S$ besitze für $v,w \in S$ die Eigenschaft

(1) $$Tv \leq Tw \Rightarrow v \leq w.$$

Dann heißt das Problem

(2) $$Tu = r$$

bei gegebener rechter Seite $r \in S$ "von monotoner Art". Synonym dazu nennt man heute den Operator T "invers-isoton", falls er (1) befriedigt (siehe etwa J.Schröder [8]).

Die Untersuchung solcher Aufgaben erwies sich im vergangenen Jahrzehnt sowohl theoretisch als auch praktisch als außerordentlich fruchtbar, man vergleiche dazu z.B. das Buch von W.Walter [9] über Differential- und Integral-Ungleichungen oder die Publikationen von J.Schröder (siehe etwa [8] und die dort zitierte Literatur). Schaut man in neuere Publikationen oder z.B. in Bücher über Funktional-Analysis, so muß man leider trotzdem erkennen:

Die invers-isotonen Operatoren werden trotz ihrer großen und unbestreitbaren Erfolge in der Mathematik immer noch nicht allgemein als selbstverständliches Hilfsmittel benutzt.

Wörtlich dasselbe gilt auch für die Intervall-Mathematik. Dieses Wort wurde von L.Fox im Jahre 1974 geprägt als Oberbegriff für Intervall-Arithmetik, Intervall-Analysis, Intervall-Algebra, ... Inzwischen gibt es auf diesem Gebiet mehr als 600 Originalarbeiten; zwei Bibliographien wurden von F.Bierbaum [3],[4] herausgegeben.

In der folgenden Note soll erstmals ausführlich an Hand einfacher Beispiele gezeigt werden, daß invers-isotone Operatoren und Intervall-Analysis eng zusammenhängen und gemeinsam mit Erfolg zur konstruktiven Lösung von Problemen der Analysis benutzt werden können. Auf der GAMM-Tagung 1976 in Graz wurde schon kurz darüber berichtet [7].

2. Bezeichnungen und Aufgabenstellung

Die Menge S mit den Elementen a,b ..., u,v,..., sei halbgeordnet mit der Ordnungsrelation $\leq$. Die Elemente der Potenzmenge $\mathbb{P}(S)$ sollen mit $\{a\},\ldots,$ $\{u\}$, ... bezeichnet werden. Wie üblich sei $\mathbb{I}(S)$ die Menge aller Intervalle $[a] = [\underline{a},\overline{a}] := \{x \in S \mid \underline{a} \leq x \leq \overline{a}\}$ mit $\underline{a},\overline{a} \in S, \underline{a} \leq \overline{a}$. Es sei $\{u\} \in \mathbb{P}(S)$ und in S existiere inf $\{u\}$, sup $\{u\}$ (das ist z.B. der Fall, wenn S ein vollständiger Verband ist). Man setzt dann

$$\text{Intv}\{u\} := [\inf\{u\}, \sup\{u\}]$$

und nennt dieses Intervall "Intervall-Hülle" von $\{u\}$ bzw. (nach Nuding) "optimale Intervall-Einschließung" von $\{u\}$. Eine Menge von Operatoren $T : S \to S$ wird entsprechend als $\{T\}$ bezeichnet. Damit kann man anstatt der einen Operator-Gleichung (2) eine Menge von Operatorgleichungen

$$(3) \qquad \{T\}\, u = \{r\}$$

betrachten. Ihre Lösungsmenge werde stets mit

$$\{\hat{u}\} := \{u \in S \mid Tu = r, \quad T \in \{T\}, \quad r \in \{r\}\}$$

bezeichnet. Diese Lösungsmenge ist im allgemeinen (selbst bei "einfachen" Problemen) nicht auf einfache Weise darstellbar oder charakterisierbar. Daher ist man an einfachen Einschließungen von $\{\hat{u}\}$ interessiert. Die optimale Intervalleinschließung von $\{\hat{u}\}$ ist natürlich $[u^*] :=$ Intv $\{u\}$, jedoch ist $[u^*]$ i.a. ebensowenig wie $\{\hat{u}\}$ auf einfache Weise angebbar. Damit lassen sich die beiden folgenden Aufgaben formulieren:

Hauptaufgabe der konstruktiven Intervall-Analysis:

Zur Lösungsmenge $\{\hat{u}\}$ der Operatorgleichungen (3) ist konstruktiv eine (möglichst günstige) Intervalleinschließung $[\hat{u}]$ anzugeben, für die also gilt

$$\{\hat{u}\} \subseteq [u^*] := \mathrm{Intv}\,\{\hat{u}\} \subseteq [\hat{u}]. \tag{4}$$

Diese Hauptaufgabe wird von den gängigen Intervall-Verfahren gelöst. Die Einschließungen durch $[\hat{u}]$ sind in vielen Fällen recht günstig, allerdings gibt es teilweise auch pessimistische Abschätzungen. Verlangt man Optimalität der Einschließung, so kann man etwa fordern:

Optimalisierungsaufgabe der Intervall-Analysis:

Es sind Klassen von Problemen (3) zu kennzeichnen, für die die Einschließung (4) optimal ist, d.h. für die

$$\underline{u}^*, \overline{u}^* \in \{\hat{u}\} \subseteq [u^*] = [\underline{u}^*, \overline{u}^*] := \mathrm{Intv}\,\{\hat{u}\} \tag{5}$$

gilt.

Weiter sind zu diesen Klassen noch konstruktive Verfahren anzugeben, durch die $[u^*]$ (i.a. näherungsweise) erzeugt wird.

Dies gibt Anlaß zu der folgenden

Definition: Ein Verfahren zur Bestimmung einer Intervalleinschließung $[\hat{u}]$ für die Lösungsmenge $\{\hat{u}\}$ der Operatorgleichungen (3) heiße schrankentreu, wenn es die optimale Intervalleinschließung $[u^*]$ liefert, d.h. wenn gilt

$$[\hat{u}] = \mathrm{Intv}\,\{\hat{u}\} \tag{6}$$

In der nachfolgenden Arbeit werden einige Problemklassen angegeben, in denen die Optimalität nach (5) gewährleistet ist, und/oder für die die Schrankentreue nach (6) nachgewiesen werden kann. In den meisten Fällen läßt sich dabei $[u^*]$ sogar explizit angeben oder wenigstens kennzeichnen. Die beiden Bilder geben eine Veranschaulichung der drei Mengen $\{\hat{u}\}$, $[u^*] := \mathrm{Intv}\,\{\hat{u}\} = [\inf \{\hat{u}\}, \sup \{\hat{u}\}]$, $[\hat{u}]$. Es ist $S := \mathbb{R}^2$ gewählt, die Ordnungsrelation $\leq$ ist (wie üblich) komponentenweise betrachtet.

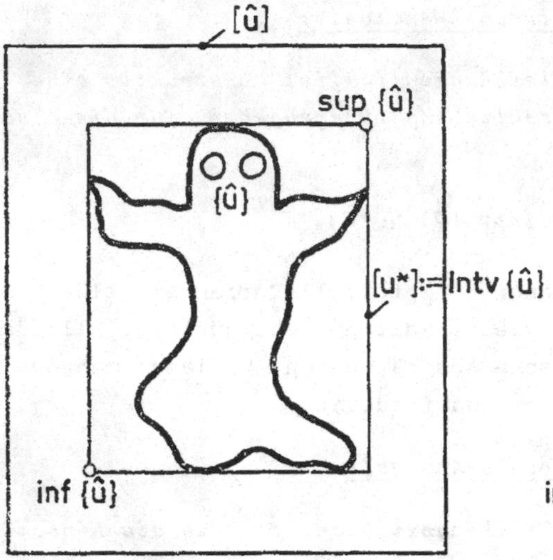

Bild 1

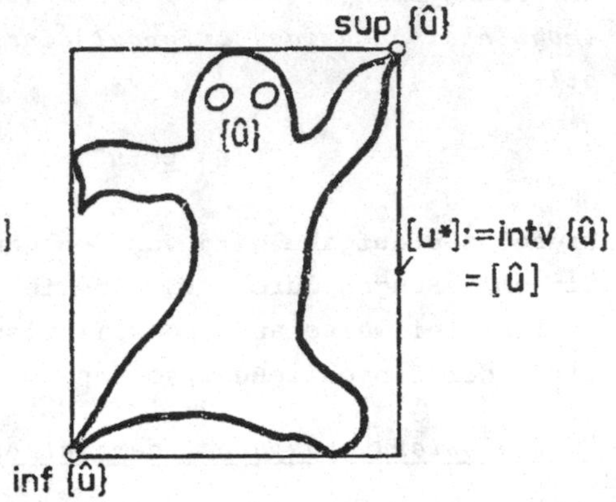

Bild 2

Bild 1: Allgemeine Lage der drei Mengen gemäß (4).

Bild 2: Sonderfall der Optimalität (5) und der Schrankentreue (6).

3. Zwei einfache Beispiele

Im Lösungsraum der Operatorgleichung (2) besteht die Menge $\{\hat{u}\}$ der Lösungen nur dann aus einem einzigen Element $\hat{u}$, wenn (2) eindeutig lösbar ist. Die "meisten" Probleme der Mathematik sind jedoch nicht eindeutig lösbar und damit nicht "sachgemäß gestellt" (auch wenn sie aus den Anwendungen kommen). Ein typisches Beispiel dafür ist das Anfangswertproblem

$$u'(t) = 2\sqrt{u(t)}\ , \quad u(o) = o. \tag{7}$$

Im Raum $C^1[o,\infty)$ hat die allgemeine Lösung von (7) die Gestalt

$$\hat{u}(t) = \begin{cases} o & \text{für } o \leq t \leq t_o\,, \\ (t - t_o)^2 & \text{für } t_o < t\,. \end{cases}$$

Die Lösungsmenge $\{\hat{u}\}$ liegt in dem Funktionsintervall $[u^*] = [o, t^2]$. Man beachte jedoch, daß nicht jede Funktion $v \in [u^*]$ Lösung von (7) ist, d.h. es ist $\{\hat{u}\} \neq \text{Intv}\,\{\hat{u}\}$. Jedoch ist $[u^*]$ optimal im Sinne von (6), denn $\underline{u}^*(t) := o$ und $\overline{u}^*(t) := t^2$ lösen (7).

In fast allen Anwendungsaufgaben treten Eingangsdaten auf, die z.B. aus physikalischen Messungen gewonnen sind und mithin prinzipiell fehlerbehaftet und nie "exakt" angebbar sind. Ein typisches Beispiel dafür ist der Fall mit Luftwiderstand. Das dazugehörige Anfangswertproblem lautet

$$(8)\qquad \begin{cases} u''(t) = g - \frac{\rho}{2} c_w \frac{F}{m} u'(t)^2, \\ u(o) = u'(o) = o. \end{cases}$$

Darin bedeutet u die Fallstrecke, t die Zeit, g die Erdbeschleunigung, ρ die Luftdichte, m die Körpermasse, c_w den Luftwiderstandsbeiwert und F die Körperfläche. Die Lösung von (8) ist eindeutig bestimmt und lautet

$$(9)\qquad \hat{u}(t) = a^{-2} \ln \cosh a\sqrt{g}\, t$$

mit

$$a := \sqrt{\frac{\rho}{2} c_w \frac{F}{m}} .$$

Nun ist aber z.B. $g = 9.807 \pm 0.027 \ m/s^2 \in [9.780, 9.834] \ m/s^2$ und auch die in a auftretenden Größen sind gemessene Werte, also nur mit mäßiger Genauigkeit angebbar. Damit sind in (9) die Konstanten a und g durch gewisse Datenintervalle $[a] = [\underline{a}, \overline{a}]$ und $[g] = [\underline{g}, \overline{g}]$ zu ersetzen; an die Stelle einer Lösung (9) tritt also die Lösungsmenge

$$(10)\qquad \{\hat{u}(t)\} := [a]^{-2} \ln \cosh [a] \sqrt{[g]}\, t.$$

Die Funktionen $\hat{u}(t)$ nach (9) ist bei festem Wert t isoton bezüglich g und antiton bezüglich a. Also hat die Intervallhülle $[u^*] = [\underline{u}^*, \overline{u}^*] :=$ Intv $\{\hat{u}\}$ zu $\{\hat{u}\}$ die Grenzen

$$\underline{u}^*(t) := \overline{a}^{-2} \ln \cosh \overline{a} \sqrt{\underline{g}}\, t ,$$

$$\overline{u}^*(t) := \underline{a}^{-2} \ln \cosh \underline{a} \sqrt{\overline{g}}\, t .$$

Da diese Schranken selbst Lösungen der Differentialgleichung und der Anfangsbedingungen (8) sind, ist $[u^*]$ optimal im Sinne von (5).

Berechnet man zu (10) nach den Regeln der Intervallarithmetik (siehe R.E.Moore [6]) eine einschließende Intervallfunktion, so erhält man etwa $[\hat{u}] = [\underline{u}, \overline{u}]$ mit

$$\begin{cases} \underline{u}(t) := [a]^{-2} \times \ln\,(\exp \underline{a}\,\sqrt{\underline{g}}\; t + \exp -\overline{a}\sqrt{\overline{g}}\; t)/2, \\ \overline{u}(t) := [a]^{-2} \times \ln\,(\exp \overline{a}\,\sqrt{\overline{g}}\; t - \exp -\underline{a}\sqrt{\underline{g}}\; t)/2. \end{cases}$$

Ein <u>Zahlenbeispiel</u> ist : $\underline{a} := 1/2$, $\overline{a} := 2$, $\sqrt{[g]}\; t := 1$. Dies liefert $[u^*] = [0.331\ldots, 0.480\ldots]$, $[\hat{u}] = [-0.457\ldots\,,\ 5.542\ldots]$.Der Quotient der Spannen ist 40.212..., d.h. das optimale Intervall $[u^*]$ wird durch die Intervalleinschließung $[\hat{u}]$ um mehr als den Faktor 40 überschätzt!!! Dabei ist sogar $0 \in [\hat{u}]$, obwohl nach (10) sicher $\{\hat{u}(t)\} > 0$ ist.

Man könnte vermuten, daß dieses pessimistische Ergebnis von der Zerspaltung der Funktion cosh in die Exponentialfunktion herrührt. Aber auch mit der ebenfalls korrekten intervallarithmetischen Darstellung

$$\underline{u}(t) := \overline{a}^{-2} \ln \cosh \underline{a}\sqrt{\underline{g}}\; t\ ,$$

$$\overline{u}(t) := \underline{a}^{-2} \ln \cosh \overline{a}\sqrt{\overline{g}}\; t$$

erhält man im vorliegenden Zahlenbeispiel mit $[\hat{u}] = [0.003\ldots, 5.300\ \ldots]$ immer noch eine Überschätzung um mehr als den Faktor 35!

Die Ergebnisse der beiden voranstehenden Beispiele sind fast selbstverständlich, da die Lösung explizit bekannt ist. Im Gegensatz dazu werden im folgenden Sätze angegeben werden für große Klassen von Gleichungen, deren Lösungen nicht explizit bekannt sind.

4. <u>Ein allgemeiner Satz</u>

Es sei S ein linearer und halbgeordneter Raum und $D \subseteq S$. Der Operator $T : S \to S$ sei additiv zerlegt als $T := L - R$, wobei $L : D \to S$ ein (festgehaltener) linearer Operator ist. Für den Restoperator $R : S \to S$ gelte $R \in \{R\}$.

<u>*Satz 1:*</u> *In $\{r\}$ gebe es zwei Elemente $\underline{r}$, $\overline{r} \in \{r\}$ derart, daß $\underline{r} \leq r \leq \overline{r}$ ist für alle $r \in \{r\}$. Es gebe zwei Operatoren $\underline{R}$, $\overline{R} \in \{R\}$ derart, daß für jedes $v \in S$ gilt: $\underline{R}v \leq Rv \leq \overline{R}v$ für alle $R \in \{R\}$. Die beiden Operatoren $\overline{T} := L - \overline{R}$, $\underline{T} := L - \underline{R}$ seien auf D inversisoton. Ferner habe die Aufgabe $\overline{T}\,u = \overline{r}$ eine größte Lösung*) $\overline{u}$ und $\underline{T}\,u = \underline{r}$ eine kleinste Lösung $\underline{u}$. Dann gilt für die Lösungsmenge $\{\hat{u}\}$ von (3) die optimale Einschließungsaussage (5).*

<u>Beweis</u>: Die Eigenschaft $\underline{u}^*, \overline{u}^* \in \{\hat{u}\}$ folgt aus der Definition. Es sei

$T\hat{u} = r$, dann ist $L\hat{u} = R\hat{u} + r \leq \overline{R}\hat{u} + \overline{r}$, also gilt $\overline{T}\hat{u} \leq \overline{r}$. Für jede Lösung v von $\overline{T}u = \overline{r}$ gilt dann wegen der Inversisotonie $\hat{u} \leq v$. Dies gilt erst recht für die größte derartige Lösung $\overline{u}$. Mithin ist $\hat{u} \leq \overline{u}$. Die andere Hälfte wird genauso bewiesen.

Bemerkung: Um optimale Schranken für die Lösungsmenge der Operatorengleichungen (3) anzugeben, sind also allein zwei spezielle Gleichungen der Gestalt (2) aufzulösen.

5. Lineare Gleichungssysteme

Es sei $S := \mathbb{R}^n$; u und r seien n-Vektoren; T sei eine $n \times n$-Matrix. Die Ordnungsrelation $\leq$ und die Inklusion $\in$ werde auf S komponentenweise erklärt. Damit lassen sich Intervall-Vektoren $[r] = [\underline{r}, \overline{r}]$ und Intervall-Matrizen $[T] = [\underline{T}, \overline{T}]$ definieren. In der Vektormenge $\{r\}$ gebe es zwei Schrankenelemente $\underline{r}, \overline{r} \in \{r\} \subseteq [\underline{r}, \overline{r}]$. Analog mögen in der Matrizenmenge $\{T\}$ zwei Matrizen $\underline{T}, \overline{T}$ existieren mit $\underline{T}, \overline{T} \in \{T\} \subseteq [\underline{T}, \overline{T}]$. Damit wird (3) zu einer Menge von linearen Gleichungssystemen. Die (als eindeutig vorausgesetzte) Lösung von $Tu = r$ werde mit $\hat{u}(T,r)$ bezeichnet. Eine Matrix T heißt inverspositiv, wenn T^{-1} existiert und (komponentenweise) $T^{-1} \geq o$ ist.

Satz 2: Die beiden Matrizen $\underline{T}, \overline{T}$ seien inverspositiv. Dann lassen sich in den drei folgenden Fällen +) *die optimalen Intervalleinschließungen $[u^*] := \mathrm{Intv}\,\{\hat{u}\}$ zur Lösungsmenge $\{\hat{u}\}$ von (3) sogar explizit angeben und es gilt (5):*

a) $[r] \geq o : [u^*] = [\hat{u}(\overline{T}, \underline{r}),\ \hat{u}(\underline{T}, \overline{r})]$,

b) $[r] \leq o : [u^*] = [\hat{u}(\underline{T}, \underline{r}),\ \hat{u}(\overline{T}, \overline{r})]$,

c) $[r] \ni o : [u^*] = [\hat{u}(\underline{T}, \underline{r}),\ \hat{u}(\underline{T}, \overline{r})]$.

---

*) Dieser Begriff ist nicht zu verwechseln mit "maximaler" Lösung gemäß der (üblichen)
Definition: Die Lösung $\overline{u}$ von $Tu = r$ heißt größte Lösung, wenn für jede Lösung $\hat{u}$ stets $\hat{u} \leq \overline{u}$ gilt. Eine Lösung u^* heißt maximal, wenn aus $\hat{u} \geq u^*$ folgt, daß $\hat{u} = u^*$ ist.

+) wodurch jedoch nicht alle möglichen Kombinationen erledigt sind.

Beweis: Da die Matrizen $\underline{T},\overline{T}$ inverspositiv sind, gilt für die Inversen zu $T \in \{T\} \subseteq [\underline{T},\overline{T}]$ stets $o \leq \overline{T}^{-1} \leq T^{-1} \leq \underline{T}^{-1}$. Damit folgt etwa für $[r] \geq o$ aus $\underline{r} \leq T\hat{u} \leq \overline{r}$ die Ungleichungskette $\hat{u}(\overline{T},\underline{r}) = \overline{T}^{-1}\underline{r} \leq T^{-1}\underline{r} \leq T^{-1}r = \hat{u} \leq T^{-1}\overline{r} \leq \underline{T}^{-1}\overline{r} = \hat{u}(\underline{T},\overline{r})$. Die anderen Fälle folgen analog.

Eine Matrix $T = (t_{ik})$ heißt M-Matrix, wenn $t_{ik} \leq o$ für $i \neq k$ ist und T^{-1} existiert mit $T^{-1} \geq o$. Jede M-Matrix ist daher inverspositiv. Eine Intervall-Matrix $[T] = [\underline{T},\overline{T}]$ heißt M-Matrix-Intervall, wenn jede Matrix $T \in [T]$ eine M-Matrix ist. Dies ist der Fall, wenn $\underline{T}$ eine M-Matrix ist und wenn $\overline{t}_{ik} \leq o$ für $i \neq k$ gilt.

Satz 3: *(Barth [1], Beeck [2], Nuding [1]): Ist [T] ein M-Matrix-Intervall, so liefert das Intervall-Gauß-Verfahren, angewandt auf die linearen Intervallgleichungssysteme $[T]\,u = [r]$ in den drei Fällen a), b), c) von Satz 2 das optimale Ergebnis $[u^*]$, d.h. dieses Gaußsche Eliminationsverfahren ist schrankentreu.*

Es sei $\rho(T)$ der Spektralradius der Matrix T. Entsprechend zum Vorgehen bei Satz 1 wird nun T aufgespalten in $T = I - R$ mit der Identität I. Es gilt

Satz 4: *(Barth [1]): In der Intervallmatrix $[R] = [\underline{R},\overline{R}]$ sei $\underline{R} \geq o$ und $\rho(\overline{R}) < 1$. Man betrachtet die linearen Gleichungssysteme*

$$u = [R]\,u + [r]. \tag{11}$$

Dann konvergiert das Iterationsverfahren, ausgehend von einem beliebigen Anfangsvektor, gegen die optimale Intervalleinschließung Intv $\{\hat{u}\}$ der Lösungsmenge $\{\hat{u}\}$ von (11), ist also schrankentreu.

Bemerkungen: Man zeigt leicht, daß die Matrizen $[T] := I - [R]$ in Satz 4 alle M-Matrizen sind. Trotzdem geht Satz 4 über die Ergebnisse von Satz 3 und Satz 2 hinaus, weil dabei beliebige rechte Seiten $[r]$ zugelassen sind!

6. Anfangswertprobleme bei gewöhnlichen Differentialgleichungen.

Betrachtet werden die Lösungsmengen $\{\hat{u}\} = \{\hat{u}(t)\}$ der Anfangswertprobleme

$$(12)\quad \begin{cases} u'(t) = f(t,u(t)), \\ u(o) = \eta. \end{cases}$$

Dabei seien η, $u = (u_i(t))$, $f = f_i(t,x)$ n-Vektoren mit $\eta \in \mathbb{R}^n$, $u_i \in C^1[o,T]$, $f_i : [o,T] \times \mathbb{R}^n \to \mathbb{R}$. Als Ordnungsrelation $\leq$ wird die punktweise Relation verwendet. Es ist also z.B. $u \leq \overline{u}$, falls $u_i(t) \leq \overline{u}_i(t)$ gilt für alle $i = 1(1)n$ und alle $t \in [o,T]$. Entsprechend ist $f \leq \overline{f}$, wenn $f_i(t,x) \leq \overline{f}_i(t,x)$ gilt für alle $i = 1(1)n$, alle $t \in [o,T]$ und alle Vektoren $x \in \mathbb{R}^n$. In der Bezeichnungsweise von Satz 1 ist $L\,u \mathrel{\hat{=}} du(t)/dt$ und $Ru \mathrel{\hat{=}} f(t,u(t))$. Die Funktion $f(t,x)$ heiße entsprechend zu W.Walter [9] quasiisoton (bei Walter quasimonoton), falls alle $f_i(t,x)$ isoton sind bezüglich allen Variablen x_k mit $k \neq i$.

Satz 5: *Es sei* $\underline{\eta}$, $\overline{\eta} \in \{\eta\} \subseteq [\underline{\eta},\overline{\eta}]$; $\underline{f}$, $\overline{f} \in \{f\} \subseteq [\underline{f},\overline{f}]$. *Die Funktionen* $\underline{f}$, $\overline{f}$ *seien quasiisoton mit* $\underline{f}_i, \overline{f}_i \in C^o[o,T] \times \mathbb{R}^n$. *Die Maximallösung von* $u' = \overline{f}(t,u)$, $u(o) = \overline{\eta}$ *werde mit* $\overline{u}$, *die Minimallösung von* $u' = \underline{f}(t,u)$, $u(o) = \underline{\eta}$ *mit* $\underline{u}$ *bezeichnet. Dann gilt für die Menge* $\{\hat{u}(t)\}$ *der Lösungen von*

$$(13)\quad \begin{cases} u' = \{f(t,u)\}, \\ u(o) = \{\eta\} \end{cases}$$

-soweit sie in $[o,T]$ *existieren - die Optimalitätsaussage (5).*

Beweis: Die folgenden Überlegungen brauchen nicht im ganzen Intervall $[o,T]$ zu gelten, sie sind jeweils in demjenigen größten Teilintervall von $[o,T]$ durchzuführen, in dem die betrachteten Lösungen existieren. Für jede Lösung $\hat{u}$ von (12) gilt $\hat{u}' = f(t,\hat{u}) \leq \overline{f}(t,\hat{u})$, $\hat{u}(o) = \eta \leq \overline{\eta}$. Die Maximallösung $\overline{u}$ existiert nach bekannten Sätzen (vgl. W.Walter [9]) und läßt sich von oben beliebig genau durch eine Funktion $v \in C^1[o,T]$ approximieren, die Lösung der Differentialungleichung $v' > \overline{f}(t,v)$, $v(o) > \overline{\eta}$ ist. Nach der Theorie der Differential-Ungleichungen (vgl. W.Walter [9]) ist dann $\hat{u} < v$ und damit in der Grenze $\hat{u} \leq \overline{u}$ wie behauptet. Die andere Ungleichung wird genau so bewiesen.

Bemerkungen: 1) Anstelle mit der Schreibweise in (13) lassen sich die betrachteten Anfangswertaufgaben auch in der Gestalt

$$u' \in \{f(t,u)\},$$
$$u(o) \in \{\eta\}$$

schreiben. Diese Schreibweise ist in der Volkswirtschaft, der Kontrolltheorie, ... üblich geworden, bedeutet aber offensichtlich denselben Sachverhalt, der hier betrachtet wird.

2) Man beachte, daß die Stetigkeit nur für die Schrankenfunktionen $\underline{f},\overline{f}$ gefordert wird, nicht aber für die dazwischen liegenden Funktionen f. Möglicherweise besitzen damit die "Zwischenprobleme" keine oder keine eindeutig bestimmte Lösung. Für jede existierende Lösung $\hat{u}(t)$ gilt trotzdem die Aussage des Satzes!

7. Randwertprobleme bei gewöhnlichen Differentialgleichungen.

Man betrachtet in $a \leq t \leq b$ die speziellen linearen Differentialgleichungen zweiter Ordnung

$$(14) \qquad -u''(t) = f(t)\, u(t) + g(t)$$

unter den Randbedingungen erster Art

$$(15) \qquad u(a) = \alpha, \; u(b) = \beta.$$

Es seien $f,g \in C[a,b]$, sowie $f(t) \geq o$ auf $[a,b]$. Dann hat bekanntlich das Randwertproblem (14), (15) eine eindeutig bestimmte Lösung aus $C^2[a,b]$ sie werde mit $\hat{u}(t;f,g,\alpha,\beta)$ bezeichnet. Da für $f \geq o$ die Aufgabe (14), (15) von monotoner Art ist, erhält man durch "Aufblähen" von f,g,α,β zu Mengen sofort den

Satz 6: Die reellen Punktmengen $\{\alpha\},\{\beta\}$ mögen die folgenden Eigenschaften besitzen: $\underline{\alpha},\overline{\alpha} \in \{\alpha\} \subseteq [\underline{\alpha}\ \overline{\alpha}]$; $\underline{\beta},\overline{\beta} \in \{\beta\} \subseteq [\underline{\beta},\overline{\beta}]$. Entsprechend seien $\{f(t)\},\{g(t)\}$ zwei Funktionsmengen mit $\underline{f},\overline{f} \in \{f\} \subseteq [\underline{f},\overline{f}]$; $\underline{g},\overline{g} \in \{g\} \subseteq$ $\subseteq [\underline{g},\overline{g}]$. Die vier Schrankenfunktionen $\underline{f},\overline{f},\underline{g},\overline{g}$ seien aus $C[a,b]$, ferner gelte $\underline{f} \geq o$ auf $[a,b]$. Dann läßt sich die Lösungsmenge $\{\hat{u}(t)\}$ zu den Randwertproblemen

$$-u''(t) = \{f(t)\}u(t) + \{g(t)\},$$

$$u(a) = \{\alpha\}, \; u(b) = \{\beta\}$$

in den folgenden drei Fällen im Sinne von (5) optimal einschließen:

a) $[g],[\alpha],[\beta] \geq 0 : [u^*] = [\hat{u}(t;\underline{f},\underline{g},\underline{\alpha},\underline{\beta}),\ \hat{u}(t;\overline{f},\overline{g},\overline{\alpha},\overline{\beta})]$,

b) $[g],[\alpha],[\beta] \leq 0 : [u^*] = [\hat{u}(t;\overline{f},\underline{g},\underline{\alpha},\underline{\beta}),\ \hat{u}(t;\underline{f},\overline{g},\overline{\alpha},\overline{\beta})]$,

c) $[g],[\alpha],[\beta] \ni 0 : [u^*] = [u(t;\overline{f},\underline{g},\underline{\alpha},\underline{\beta}),\ \hat{u}(t;\overline{f},\overline{g},\overline{\alpha},\overline{\beta})]$.

Beweis: Man setzt in Satz 1 für $D := C^2[a,b] \times \mathbb{R} \times \mathbb{R}$, $S := C[a,b] \times \mathbb{R} \times \mathbb{R}$. Der Operator Lu bestehe aus dem Vektor $(u''(t),\ u(a),\ u(b)$, während der Operator Ru durch den Vektor $(f(t)u(t) + g(t),\alpha,\beta)$ ersetzt wird. Den drei Fällen a),b),c) entsprechen dann die drei Vorzeichenverteilungen $\hat{u} \geq 0$, $\hat{u} \leq 0$, $0 \in \hat{u}$. Die Lösungen $\hat{u}$ sind in allen drei Fällen isoton bezüglich g,α,β. Bezüglich f sind sie bei a) isoton und bei b) antiton. Das Ergebnis bei c) entsteht durch Kombination von a) und b).

Bemerkungen: 1) Die Übertragung von Satz 6 auf allgemeinere Randwertprobleme (Auftreten von u' und/oder Nichtlinearitäten in der rechten Seite von (14), Randbedingungen von zweiter oder dritter Art in (15) bzw. nichtlineare Randbedingungen) ist offensichtlich sofort möglich, soweit dadurch die Inversisotonie nicht zerstört wird.

2) Die linken Seiten in den drei Fällen a),b),c) sind als Kurzschrift aufzufassen für $\underline{g}(t) \geq 0$ in $[a,b]$, $\underline{\alpha} \geq 0, \ldots, \underline{g}(t) \leq 0 \leq \overline{g}(t)$ in $[a,b]$, $\underline{\alpha} \leq 0 \leq \overline{\alpha}$, $\underline{\beta} \leq 0 \leq \overline{\beta}$.

3) Man beachte, daß die Stetigkeit nur für die Schrankenfunktionen $\underline{f},\overline{f},\underline{g},\overline{g}$ gefordert wird, nicht aber für die dazwischen liegenden Funktionen f,g. Möglicherweise besitzen damit die "Zwischenprobleme" keine oder keine eindeutig bestimmte Lösung. Für jede existierende Lösung $\hat{u}(t)$ gilt trotzdem die Aussage des Satzes.

8. Ausblick

In den Kapiteln 5 bis 7 wird an einfachen Beispielen gezeigt, wie sich der allgemeine Satz des Kapitels 4 in speziellen Fällen realisieren läßt, außerdem werden noch einige Erweiterungen angegeben. Bis heute sind schon sehr viele Operatoren als inversisoton nachgewiesen; man denke etwa an Systeme von Volterra'schen Integralgleichungen; parabolische, elliptische, hyperbolische Randwertprobleme etc. Mit dem Satz 1 von Kapitel 4 lassen sich damit die optimalen Einschließungen für die Lösungsmengen von entsprechenden Mengen von Operatorgleichungen angeben. Fast selbstverständlich ist z.B. ein Satz über die Lösung $\hat{u}(t,x)$ der parabolischen Differentialgleichung

$$u_t = \{f(t,x,u,u_x,u_{xx}\}$$

unter passenden mengenwertigen Randbedingungen in Erweiterung von Satz 5. Weitgehend unbekannt sind jedoch noch Folgerungen von der Gestalt der Sätze 3 und 4, in denen also Aussagen gemacht werden über die Schrankentreue von speziellen (Iterations-) Verfahren.

Anschrift des Verfasser:

Karl Nickel
Institut für Angewandte Mathematik
der Albert-Ludwigs-Universität Freiburg
Hermann-Herder-Str. 10
7800 Freiburg i.Br.

9. Schrifttum:

[1] Barth,W. und E.Nuding: Optimale Lösung von Intervallgleichungssystemen. Computing 12 (1974), 117-125.

[2] Beeck,H.: Zur scharfen Außenabschätzung der Lösungsmenge bei linearen Intervallgleichungssystemen. ZAMM 54 (1974), T 208 - T 209.

[3] Bierbaum,F.: Intervall-Mathematik. Eine Literaturübersicht. Interner Bericht Nr. 74/2 des Instituts für Praktische Mathematik der Universität Karlsruhe, (1974).

[4] Bierbaum,F.: Intervall-Mathematik. Eine Literaturübersicht. Nachtrag. Interner Bericht Nr. 75/3 des Instituts für Praktische Mathematik der Universität Karlsruhe, (1975).

[5] Collatz,L.: Aufgaben monotoner Art. Arch.Math. 3 (1952), 366-376.

[6] Moore,R.E.: Intervallanalyse. R.Oldenbourg (1969).

[7] Nickel,K.: Aufgaben monotoner Art und Intervall-Mathematik. Erscheint in der ZAMM.

[8] Schröder,J.: Upper and Lower Bounds for Solutions of Generalized Two-Point Boundary Value Problems. Numer.Math. 23 (1975), 433-457.

[9] Walter,W.: Differential and integral inequalities. Springer (1970).

THE RKFHB4 METHOD FOR DELAY - DIFFERENTIAL EQUATIONS

Jesper Oppelstrup
Numerical Analysis
Royal Institute of Technology
S-100 44 STOCKHOLM, Sweden

ABSTRACT

A method for the numerical solution of initial value problems for systems of retarded delay-differential equations is described. It uses a variable step Runge-Kutta Fehlberg 4/5 method combined with fourth degree Hermite-Birkhoff interpolation, which makes the usual local error estimator asymptotically correct. To obtain good numerical stability, the basic interpolation is modified for small delays. The possible initial discontinuities are algorithmically treated by using a very short stepsize in the critical step, and to this end the stepsize control has been somewhat modified. Numerical examples are included.

INTRODUCTION

In recent years, there has been growing interest in models for dynamical systems where the rate of change depends on the whole history of the system and not only, as in ordinary differential equations, on the present state vector. Such models arise in many fields of applications, [13].

The method described here is applicable to initial value problems for systems of retarded delay-differential equations on a finite interval $[t_0,T]$

$$\begin{cases} \dot{y}(t) = f(t,y(t),\, y(t-\tau_1),\, \dots,\, y(t-\tau_m)), \; t \geq t_0 \\ y(s) = g(s), \; s \leq t_0 \\ \tau_i = \tau_i(t), \;\; i = 1,2,\dots,m \end{cases} \tag{1}$$

where $y,f \in R^n$ and $\tau_i \geq 0$, hence the term retarded. Dots denote differentiation w.r.t. t, $\dot{} = \frac{d}{dt}$. As an example, consider the single equation with a constant delay

$$\begin{cases} \dot{y}(t) = q\,y(t-\tau) & t \geq 0 \\ y(s) = \quad 1 & s \leq 0 \end{cases}$$

which, for $q = -1.5$, $\tau = 1$, has the solution graphed below.

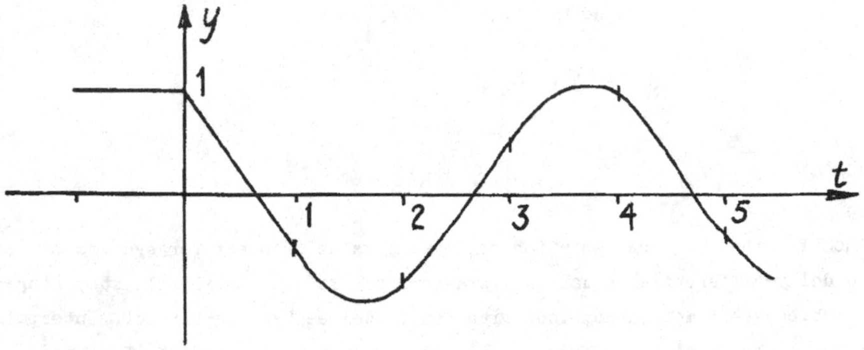

It exemplifies the smoothing property of retarded equations: the jump in $\dot{y}$ at $t = 0$ is inherited as a jump in $\ddot{y}$ at $t = \tau$, and generally, $y^{(k)}$ is discontinuous at $t = (k-1)\tau$, $k = 1,2,\ldots$. Only for special initial functions g the solution is smooth for all $t \geq 0$.

Because of these discontinuities, much work has concentrated on low-order methods such as generalizations of Euler's method, [2], [1]. For problems with smooth solutions, higher order methods have been studied. The class of problems

$$\dot{y}(t) = f(t,y(t),y(\alpha(t)))$$
$$y(t_0) = y_0$$

where the function α is such that $t_0 \leq \alpha(t) \leq t$ for all t in $[t_0,T]$, has been discussed by Stetter [14] and Feldstein [6]. Owing to the conditions on α there are no initial jumps. Tavernini [16] suggests one-step methods and obtains a fourth order method with six evaluations per step of the derivatives.

The use of extrapolation to obtain higher accuracy was advocated by Feldstein for the class of problems above, and for more general functional differential equations by Cooke and List, [1]. They assume that the positions and magnitudes of the jumps can be calculated in advance, and are able to correct for the jumps to make the extrapolation work. A similar idea is exploited by Zwerkina [17], who designed Adams-type multistep methods with corrections for jump discontinuities.

ONE-STEP METHODS WITH INTERPOLATION FOR VOLTERRA FUNCTIONAL DIFFERENTIAL EQUATIONS

The retarded delay-differential equation (1) is a member of the more general class of Volterra functional differential equations

$$\begin{cases} \dot{y}(t) = (Vy)(t) & t \geq t_0 \\ y(s) = g(s) & s \leq t_0 \end{cases} \tag{2}$$

Here, V is a Volterra functional, i.e. $(Vy)(t)$ depends on values of $y(r)$ for $r \leq t$ only. An existence - uniqueness theorem for the initial value problem (2) was given by Driver, [4].

THEOREM

If g is continuous for all $s \leq t_0$, V satisfies the Lipschitz condition

$$\|(Vy_1)(t) - (Vy_2)(t)\| \leq L \max_{s \leq t} \|y_1(s) - y_2(s)\|$$

for continuous functions y_1 and y_2, and $(Vy)(t)$ is continuous in t, then there is a unique, continuous solution of (2) on $[t_0, t_0 + H)$ for sufficiently small H.

To display more explicitly the dependence of $(Vy)(t)$ on $y(t)$ we shall write (2) in the form

$$\begin{cases} \dot{y}(t) = f(y(t), z(t)) & t \geq t_0 \\ z(t) = (Fy)(t) & \end{cases}$$
$$y(s) = g(s) \qquad s \leq t_0 \tag{3}$$

where we now assume that F is a Lipschitz continuous Volterra functional, and f is Lipschitz continuous in both arguments.

The numerical method generates a sequence y_i, $i = 0,1,2,\ldots$, which approximate the exact solution $y(t_i)$ on the grid

$$t_N < \ldots < t_{-1} < t_0 < t_1 < t_2 < \ldots < t_K = T$$

(K and N are integers)

We need a numerical approximation $\widetilde{F}$ to F. When the method steps from t_j to t_{j+1} we have available y_k, $k \le j$, so $(Fy)(t)$ is approximated by $(\tilde{F}Y)(t)$, where Y^T denotes the super-vector $\left(y_{-N}^T, y_{-1}^T, y_0^T, \dots, y_K^T\right)$. Now we can define a onestep method with interpolation for (3):

Apply a onestep method for ordinary differential equations to (3), now assuming that $z(t)$ is a known function of t; then we get

$$y_{n+1} = y_n + (t_{n+1} - t_n)\phi(y_n, z)$$

Here we have used the notation of Henrici [11] for the increment function ϕ. In the sequel, for brevity of notation, we assume that the grid is equidistant with step h; the results can be modified for variable stepsize in exactly the same manner as for ordinary differential equation, see [11].

We apply the onestep method with interpolation $(\phi, \tilde{F})$ to (3) by the prescription

$$\begin{cases} y_{n+1} = y_n + h\phi(y_n, \tilde{z}) + p_n, & n \ge 0 \\ \tilde{z}(t) = (\tilde{F}Y)(t) \end{cases}$$

$$y_j = g(t_j) + w_j \qquad j \le 0$$

where p_n are local perturbations arising in computation, such as rounding errors, and w_j are errors in the starting values. We then have the following convergence result:

THEOREM

If 1) $\tilde{F}$ is Lipschitz continuous in the sense

$$\|(\tilde{F}U)(t) - (\tilde{F}V)(t)\| \le \tilde{L}(d) \max_{k \le j} \|u_k - v_k\| \tag{4}$$

for $t \le t_j + d$

2) ϕ is consistent, hence convergent, for ordinary differential equations and, as $h \to 0$,

3) $w_j \to 0$

4) $p_n/h \to 0$

5) $\|(Fy)(t) - (\tilde{F}y)(t)\| \to 0$

then the method $(\phi, \tilde{F})$ converges.

Remark 1. Note that a certain amount of extrapolation is involved in the condition (4) on $\tilde{F}$; for most Runge Kutta methods we need $d = h$.

Remark 2. $\tilde{F}$ is applied to a function u rather than a sequence U by the trivial discretization $u_k := u(t_k)$.

Proof Define the exact increment function Δ by

$$\begin{cases} y(t_{j+1}) = y(t_j) + h\Delta(y(t_j),z) \\ z(t) = (Fy)(t) \end{cases}$$

Putting $e_j = y(t_j) - y_j$ we get

$$\begin{cases} e_{j+1} = e_j + h[\Delta(y(t_j),z) - \phi(y_j,\tilde{z})] - p_j, \quad j \geq 0 \\ e_j = -w_j, \ j \leq 0 \end{cases}$$

Denoting the local error of ϕ used on (4) by

$$hr'_j = h(\phi(y(t_j),z) - \Delta(y(t_j),z))$$

and the interpolation error in step j by

$$r''_j(t) = (Fy)(t) - (\tilde{F}y)(t)$$

we have, after some adding and subtracting,

$$\|e_{j+1}\| \leq \|e_j\| + h(L_1\|e_j\| + L_2 \ (\max_{t_j \leq s \leq t_j+d} \|r''_j(s)\| +$$

$$+ \tilde{L} \max_{l \leq j} \|e_l\|)) + \|p_j\| + h\|r'_j\|, \quad j \geq 0$$

$$\|e_j\| = \|w_j\|, \quad j \leq 0$$

Here, we have used the Lipschitz constants L_1 and L_2 of ϕ w.r.t. first and second arguments, resp. Thus, $\|e_j\|$ is bounded by the increasing sequence (for $j \geq 0$).

$$v_j = \|w_j\|, \quad j \leq 0$$

$$v_{j+1} = (1 + ha_1)v_j + ha_2 \max_{l \leq j} v_l + R_j, \quad j \geq 0$$

$$= (1 + ha)v_j + R_j$$

where $a_1 = L_1$, $a_2 = L_2\tilde{L}$, and R_j is an upper bound for

$$\|p_j\| + h(\|r'_j\| + L_2 \max_{t_j \leq s \leq t_j+d} \|r''_j(s)\|).$$

Thus, putting $hR = \max_j R_j$, $w = \max_{j\leq 0} \|w_j\|$, we have

$$\|e_j\| \leq R\left(e^{a(t_j-t_0)} - 1\right)/a + we^{a(t_j-t_0)}$$

which proves convergence. *QED.*

If the solution y and the function f are smooth enough we can use higher order methods ϕ and more accurate functionals $\tilde{F}$. If the local error of ϕ is $\mathcal{O}(h^{p+1})$, and the interpolation error is $\mathcal{O}(h^q)$, the global error is of order $h^{p \min q}$. Thus, the classical Runge Kutta method with third degree polynomial interpolation gives $\mathcal{O}(h^4)$ error with four evaluations per step, as noted by Stetter [14].

The error bound obtained in this way is much too pessimistic, especially for stable problems. It can be sharpened for this case by the use of logarithmic norms, as in Dahlquist [3].

ASYMPTOTIC ERROR ESTIMATES

A better error estimate can be found if there exists an asymptotic expansion in powers of h of the global error. Several workers have established the existence for special methods of one or a few terms, [8], [6].

But for equations with true delay-terms, there is no such infinite expansion if one uses polynomial interpolation. The reason for this is seen by a simple example. Consider Euler's method with stepsize h for

$$\dot{y}(t) = qy(t-\tau), \ \tau \text{ constant} \tag{5}$$

where we take as approximation to $y(t-\tau)$ the value y_k at the nearest gridpoint t_k which is less than or equal to $t-\tau$. Then, the interpolation error $r''(t,h) = y(t-\tau) - y(t_k)$ is $\mathcal{O}(h)$, so the global error is $\mathcal{O}(h)$, but

$$\lim_{h\to 0} r''(t,h)/h \ \nexists,$$

even if the solution $y(t)$ is smooth.

In the table below, we have solved (5) over [0,1] by this method with stepsizes $h = 2^{-N}$, $N = 2(1)8$, $\tau = 0.7$, and q and the initial function chosen to make $e^{-0.5t}$ the exact solution. The second column shows the global error at $t = 1$ divided by h.

N	$(\text{error}/h)*10^4$
2	574
3	197
4	-581
5	-192
6	578
7	193
8	-578

The periodicity is explained by the fact that the interpolation error is proportional to $h \cdot$ fractional part of $(\tau/h) = h \cdot (7 \cdot 2^N) \bmod 10$.

If we use higher order interpolation, however, there will be at least one term of an asymptotic expansion in powers of h of the global error. The proof is given in detail in [12], with consideration of the difference equation for the remainders, which are indicated by $\mathcal{O}(h^{p+1})$ here.

THEOREM

Assume, that the exact solution y and the function f are sufficiently differentiable. Then, if the local error of ϕ is of order $p+1$ and the interpolation error is of order q, and $q \geq p+1 \geq 2$,

$$\|y(t_j) - y_j\| = h^p e(t_j) + \mathcal{O}(h^{p+1})$$

where $e(t)$ satisfies the linear, nonhomogenous Volterra functional equation

$$\dot{e} = (V'(y))(t)e + \varphi(t), \quad t \geq t_0$$
$$e(s) = 0 \quad s \leq t_0$$

Here, $V'(y)$ is the Fréchet derivative of V along the exact solution $y(t)$, $\varphi(t)$ is the principal error function of ϕ used on (3), and we have ignored the initial errors w_j and the local perturbations p_j.

Proof We make use of Stetter's theorem [15]. For the local error = residual when $y(t)$ is substituted into the numerical scheme, we have

$$h(\Delta(y(t_j),Fy) - \phi(y(t_j),\widetilde{F}y)) = h[h^p\varphi(t_j) + \mathcal{O}(h^{p+1}) + Q_j]$$

where $$\|Q_j\| = \|\phi(y(t_j),Fy) - \phi(y(t_j),\widetilde{F}y)\| = \mathcal{O}(h^q) = \mathcal{O}(h^{p+1}),$$

thus the local error has an expansion to at least one term.
It remains to show the stability in the sense of [15] of the inhomogenous, linearized, discretized problem, i.e.,

$$\begin{cases} u_{j+1} - u_j - h\widetilde{\phi}'U = \rho_j, & j \geq 0 \\ u_j = 0 & j \leq 0 \end{cases}$$

where $\widetilde{\phi}'$ is the Fréchet derivative of $\phi(u_j,\widetilde{F}U)$ evaluated at $u_j = y(t_j)$. Under our assumptions of differentiability and Lipschitz continuity we get for $x_j = \|u_j\|$

$$\begin{cases} x_{j+1} \leq x_j + hL(h) \max\limits_{k \leq j} x_k + \|\rho_j\| , & j \geq 0 \\ x_j = 0 , & j \leq 0 \end{cases}$$

and, putting $k^* = (t-t_0)/h$, assuming this is an integer,

$$x_{k^*} \leq h^{-1} \max_{k \leq k^*} \|\rho_k\| \cdot \left(e^{L(h)(t-t_0)} - 1\right) \Big/ L(h) =$$

$$= \mathcal{O}(h^{-1}) \max_{k \leq k^*} \|\rho_k\|$$

which is the necessary bound on the solution. Here, $L(h)$ is the norm of the linear operator $\widetilde{\phi}'$. It can be bounded in terms of h, and the Lipschitz constants for f and $\widetilde{F}$, and tends to $L_1 + L_2 \widetilde{L}$ as $h \to 0$.

Note also that linearization and discretization do not, in general, commute exactly for Runge Kutta methods owing to the off-step derivative calculations. The remaining conditions for the application of Stetter's theorem are easily seen to be satisfied. *QED.*

The functional $\widetilde{F}$ can use also the approximate values of $\dot{y}$ which are calculated anyway in integrating the differential system. The theorems hold after the obvious modifications; one only has to use majorizing recursions for 2-vectors of norms instead of scalars. Thus we can use Hermite-Birkhoff interpolation, which was suggested by Fox [8], and has, compared with Newton interpolation, smaller error constant and less sensitivity to rapid variation of the distance between gridpoints.

Remark. Delay-differential equations with state-dependent delays, say

$$\dot{y}(t) = f(t, y(t), y(t-\tau))$$

$$\tau = \tau(t, y(t))$$

are not covered by the results here, because the functional is not Lipschitz continuous.

But there is still a unique, continuous solution, and under suitable assumptions on τ, and on the interpolation method, the theorems still hold. This generalization is carried through in [12].

THE RKFHB4 METHOD

Our program uses the Runge Kutta Fehlberg 4/5 method and fourth degree Hermite-

Birkhoff interpolation. We shall discuss the treatment of the initial discontinuities and the stability of the method when used on the test equation with delay

$$\dot{y}(t) = q\,y\,(t-\tau)$$

DISCONTINUITIES, STEPSIZE REGULATION

The algorithmic treatment of differential equations with discontinuities is discussed e.g. in [7] and [9]. The difficulty is to find the point, say t^*, where the switch from one smooth right hand side to another occurs. For the clever treatment of discontinuities, therefore, the differential system is assumed given in the form

$$\dot{y}(t) := \underline{\text{if}}\ \psi(t,y) \geq 0\ \underline{\text{then}}\ f_1(t,y)\ \underline{\text{else}}\ f_2(t,y)\,;$$

and the user is required to supply the function ψ explicitly. Then, some root-finding technique can be used to obtain t^* to the desired accuracy.

To save the user from this extra work, we choose instead the brute force method of letting the normal error estimator discover the jumps.

The stepsize control suspects that there is a point of discontinuity t^* in $[t_j, t_j + h']$ if h' is the second unsuccessful stepsize tried to step from t_j. To determine when t^* has been passed, we save an upper bound $\bar{t} = t_j + h'$ on t^* (we assume integration in the positive t-direction). Over the interval $[t_j, t_j + h']$ the normal error / unit step criterion is now replaced by error per step criterion with smaller tolerance, no increase in stepsize is allowed, and whenever a step to, say $t_j + h''$, fails, $\bar{t}$ is set to $t_j + h''$ and the stepsize is halved. When the point $\bar{t}$ has been passed, the stepsize is allowed to increase, and we first try a step of length h'.

For smooth solutions, the stepsize control works like for ordinary differential equations, since the $\mathcal{O}(h^4)$ - part of the global error comes from φ, and not from interpolation. However, when the error tolerance, say ε, is not very stringent, the asymptotic theory is not really valid, and therefore the stepsize is additionally monitored so that the maximal interpolation error over any step is approximately bounded by ε.

STABILITY FOR SMALL DELAYS

To discuss the stability we first give the exact definition of the basic interpolation (later to be modified), for a single equation, to avoid double indices. In the derivative calculations when the method steps from t_j to t_{j+1}, $y(t-\tau)$ is computed as $Q_j(t-\tau)$, where Q_j is the fourth degree polynomial which satisfies

$$Q_j(t_k) = y_k, \quad k = l^*-1,\ l^*,\ l^*+1$$

$$\dot{Q}_j(t_k) = d_k, \quad k = l^*,\ l^*+1$$

where $\quad t_l < t-\tau \le t_{l+1}, \quad$ and $\quad l^* = \min(l, j-1)$.

The choice of l^* instead of l gives an explicit method even for small τ. The d_k are the approximate $\dot{y}$ - values at the gridpoints, i.e., for

$$\dot{y}(t) = f(t, y(t),\ y(t-\tau)),$$

$$d_k := f(t_k,\ y_k,\ Q_{k-1}(t_k-\tau)).$$

If $\tau < h$ the intermediate stage derivative calculations imply extrapolation. This has adverse effects on the stability.

Consider the test equation with delay (5). As $\tau \to 0$, this tends to $\dot{y}(t) = q\,y(t)$. The length of the intersection of the stability region in hq - plane with the real axis is only about 0.16 for the RKFHB4 method with basic interpolation, if $\tau \ll h$. But the stability interval of the Runge Kutta Fehlberg 4/5 method is approximately $[-3, 0]$.

One way of improving the situation is to allow the approximation to $y(t-\tau)$ to use also the current value of y,

$$y(t-\tau) \approx Q_j(t-\tau) + \omega(\tau, h)(y(t) - Q_j(t)) = R_j(t, \tau, h)$$

If we choose the continuous function $\omega(\tau, h)$ such that $\omega(0, h) = 1$, then $R_j(t, 0, h) = y(t)$, and the discretization of

$$\dot{y}(t) = f(t, y(t),\ y(t-\tau)) \quad \text{for} \quad \tau = 0$$

and

$$\dot{y}(t) = f(t, y(t),\ y(t))$$

will become the same, and the stability region will be the same for $\tau \to 0$ as for the testequation without delay.

As criteria for the choice of $\omega(\tau, h)$ we have used the accuracy and stability when the method is applied to the testequation with delay. To make ω simple to compute, we choose

$$\omega = \omega(\theta), \quad \text{where} \quad \theta = \tau/h$$

and require

$$\omega(0) = 1$$

$$\omega(\theta) = 0, \quad \text{if } \theta \ge 1$$

since no extrapolation is necessary when $\theta > 1$.

ACCURACY

Let q and τ be such that $e^{\lambda t}$ is a solution of (5), thus $q = \lambda e^{\lambda\tau}$. Let $y_n = 1$, and y_{n-1}, y_{n-2}, $\dot{y}_{n-1}$, $\dot{y}_n$ have exact values.
Then

$$y_{n+1} = e^{\lambda h} + C(\lambda h)^{p+1} + \mathcal{O}(h^{p+2})$$

with $p = 4$. To calculate the error constant C, we used the symbolic manipulation system REDUCE [10]. The result is

$$C(\omega) = (1 - 5\omega - 60\omega^3)/49920$$

$|C(1)| = 1/780$ is the maximum value of $|C|$. Thus any ω between 0 and 1 is acceptable for accuracy reasons, since for the non-delayed parts of the system the error constants are of this magnitude.

STABILITY

By the stability region S of the method we mean that set of values of (qh,θ) for which the numerical solutions are bounded. S can be represented graphically by a set of stability regions in qh-plane, one for each θ-value. To make computer plots of these, one observes that the RKFHB4 method applied to the testequation gives a linear recursion

$$X_{n+1} = A(qh,\theta)X_n$$

for the vectors $X_n = (y_n, d_n, y_{n-1}, d_{n-1}, y_{n-2}, d_{n-2})^T$. A is a 6×6 complex matrix. Thus, the interior $\underline{S}$ of S is given by

$$\underline{S} = \{(qh,\theta) |\ \text{spectral radius of } A \text{ less than } 1\}$$

By plotting $\underline{S}$ instead of S we neglect some boundary points - for instance, the origin $qh = 0$! We now choose $\omega(\theta)$ to obtain the largest $\underline{S}$, by inspecting the stability plots. It turns out that the optimal ω for each θ, say ω_{opt}, is a decreasing function of θ, and that

$$\omega = (1-\theta)^3$$

is close-to-optimal.

Note that the stability region of the differential equation is bounded (except for $\tau = 0$). If $q\tau \in D$ (see the figure) all solutions are bounded. D is bounded by the curve

$$|q\tau| = \arg(q\tau) - \pi/2$$

and its mirror image in the real line

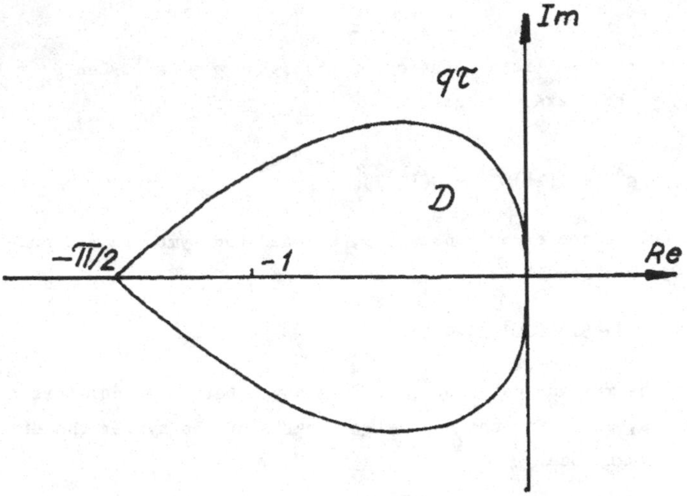

NUMERICAL EXAMPLES

The following two problems have been run with error tolerance $\varepsilon = 10^{-N}$, $N = 1(1)8$, and the maximal error e in the first component over the range and the number NFC of function evaluations were recorded.

Problem 1
$$\begin{cases} \dot{y}_1 = y_2 \\ \dot{y}_2 = -y_1/2 - 1/2 + y_1^2(t/2 - \pi/4) \end{cases}$$
$$\begin{cases} y_1(s) = \sin s \\ y_2(s) = \cos s \end{cases} \qquad s \le 2$$

over [2,20]. The exact solution is $y_1(t) = \sin t$

Problem 2
$$\begin{cases} \dot{y} = q\,y\,(t-\tau) & t \ge 0 \\ y(s) = 1 & s \le 0 \end{cases}$$

over [0,7].

$q = -1.5$, $\tau = 1$, makes the unsmooth solution barely asymptotically stable. The stability limit is $q\tau = -\pi/2$.

The exact solution is a polynomial of degree k over $[(k-1)\tau,\, k\tau]$. Putting

$$y(t) = y_k(t) = \sum_{j=0}^{k+1} a_{kj}\,(t-k\tau)^j\,, \quad t \in [k\tau,(k+1)\tau]$$
$$y_{-1}(t) = 1$$

and substituting in the differential equation, one obtains a recurrence for the a_{kj} which allows stable computation of $y(t)$.

The results are presented in log-log graphs of e^{-1} (continuous line) and ε^{-1} (broken line) versus *NFC*. Also the slope expected for a fourth order method has been indicated.

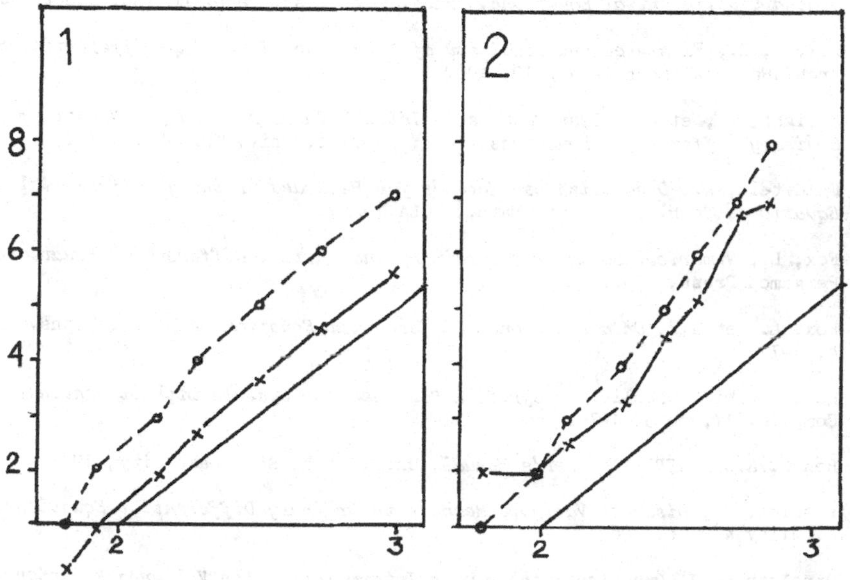

One hopes for proportionality between e and ε, the quotient e/ε being determined by the stability of the differential system and the length of the integration range. The behaviour deviates from this for small N, both because the asymptotic theory is not valid, and because the influence of the choice of starting stepsize is greater since the total number of steps is low. For the problem with discontinuities, the graph is more erratic, and it is perhaps recommendable to run such problems with different ε and compare the results.

ACKNOWLEDGEMENTS

This work was supported financially by ITM, the Swedish Institute for Applied Mathematics. It would not have been completed without professor Dahlquist's constant encouragement; he also contributed the idea of optimizing ω. The author would like to thank professor Stetter for a reference to some of his own work, of which the author was not aware.

REFERENCES

1. Cooke, K.L., List, S.E., *The Numerical Solution of Integro-Differential Equations with Retardation*, Tech.rep. 72-4, Dept. Electrical Eng., USC, Los Angeles, 1972.

2. Cryer, C.W., Tavernini, L., *The Numerical Solution of Volterra Functional Differential Equations by Euler's Method*, SIAM J. Num.Anal. 9, 1972.

3. Dahlquist, G., *Stability and Error Bounds in the Numerical Integration of Ordinary Differential Equations*, Trans. Roy. Inst. Tech. no 130, Stockholm 1959.

4. Driver, R., *Existence and Stability of Solutions of a Delay-differential System*, Arch. Rational Mech. Anal. 10, 1962.

5. Enright, W., et al., *Test Results on Initial Value Methods for Non-stiff Ordinary Differential Equations*, Tech. rep. 68, Univ. Toronto, 1974.

6. Feldstein, A., *Discretization Methods for Retarded Ordinary Differential Equations*, Tech. rep. Dept. Math., UCLA 1964.

7. Fox, L., *Numerical Solution of Ordinary and Partial Differential Equations*, Pergamon Press, 1962.

8. Fox, L., et al., *On a Functional Differential Equation*, J. Inst. Maths. Applics. 8, 1971.

9. Hay, Crosbie, Chaplin, *Integration Routines for Systems with Discontinuities*, Comp. J. 17, no 3, 1974.

10. Hearn, A.C., *REDUCE 2 User's Manual*, Univ. Utah, Salt Lake City, 1973.

11. Henrici, P., *Discrete Variable Methods in Ordinary Differential Equations*, J. Wiley & Sons, 1962.

12. Oppelstrup, J. *One-step Methods with Interpolation for Volterra Functional Differential Equations*, Tech. rep. TRITA-NA-7623, Royal Inst. Tech., Stockholm, 1976.

13. Schmitt, K. (Ed.), *Delay and Functional Differential Equations and their Applications*, Academic Press, 1972.

14. Stetter, H.J., *Numerische Lösung von Differentialgleichungen mit nacheilendem Argument*, ZAMM 45, 1965.

15. Stetter, H.J., *Asymptotic Expansions for the Error of Discretization Algorithms for Non-linear Functional Equations*, Num. Math. 7, 1966.

16. Tavernini, L., *One-step Methods for the Numerical Solution of Volterra Functional Differential Equations*, SIAM J. Num. Anal. 8, 1971.

17. Zwerkina, T.S., *Modified Adams Formula for Integrating Equations with Deviating Argument*, Univ. Druzby Narodov Patrisa Lumumby, 1, MR 32 # 2708, 1962.

Spiegelung von Stabilitätsbereichen

Rudolf Scherer

One-step methods β for the numerical solution of ordinary differential equations are studied. The corresponding stability function W_β is obtained by applying β to the scalar equation $y'=qy$, $y(o)=y_o$ $(q\in\mathbb{C})$. The stability region is defined by $S_\beta:=\{z\in\mathbb{C}\,|\,|W_\beta(z)|<1\}$. In addition to $W_\beta(z)$ we consider $\tilde{W}_\beta(z):=\{W_\beta(-z)\}^{-1}$ and to S_β the region $\tilde{S}_\beta:=\{z\in\mathbb{C}\,|\,|\tilde{W}_\beta(z)|<1\}$. The transformation from S_β to $\tilde{S}_\beta$ corresponds to the reflection $S_\beta\to\tilde{S}_\beta=\{z\in\mathbb{C}\,|\,-z\notin S_\beta\}$. To each Runge-Kutta-scheme β with generating matrix $\mathfrak{L}$ there exists the Runge-Kutta-scheme $\tilde{\beta}:=-\beta^{-1}$ with $\tilde{\mathfrak{L}}$ and with $S_{\tilde{\beta}}=\tilde{S}_\beta$ $(W_{\tilde{\beta}}=\tilde{W}_\beta)$. The generating matrix $\tilde{\mathfrak{L}}$ can be derived from $\mathfrak{L}$. The Runge-Kutta-scheme $\tilde{\beta}:=-\beta^{-1}$ is called the reflected scheme of β . The order remains the same. This reflection principle can be used to construct A-stable methods. For example the Runge-Kutta processes based on Radau and Lobatto formulae are cited.

1. Einleitung

Vorgelegt sei das Anfangswertproblem

(1) $$y'=f(y)\ ,\ y(x_o)=y_o$$

mit $f\colon G\to K^n$ $(G\subseteq K^n)$ $(K=\mathbb{R}$ bzw. $K=\mathbb{C})$ Lipschitz-stetig. Gesucht sind Näherungswerte y_{m+1} in den Gitterpunkten $x_{m+1}:=x_m+h$ $(m=o,1,\ldots,r$; Schrittweite h) für die wahre Lösung $u(x_{m+1})$.

Einschrittverfahren β lassen sich in der Form

(2) $$y_{m+1}=y_m+h\,\Phi(y_m,h)$$

schreiben, wobei die Zuwachsfunktion Φ nur von y_m und h abhängt. Bei der Anwendung von β auf die skalare *Testgleichung*

(2) $$y'=qy\ ,\ y(o)=y_o\ ,\ q\in\mathbb{C}\ ,\ \mathrm{Re}\,q<o$$

ergibt sich

(4) $$y_{m+1} = W_\beta(qh)\, y_m \quad .$$

$W_\beta(z)$ $(z:=qh)$ stellt eine rationale Approximation an die Exponentialfunktion $\exp(z)$ dar. Die Funktion W_β heißt *Stabilitätsfunktion* des Verfahrens β . Um nämlich qualitativ gleiches Verhalten der Näherungslösung und der wahren Lösung zu erreichen, wird $|W_\beta(z)| < 1$ gefordert. Dies führt auf den von DAHLQUIST [6] eingeführten Stabilitätsbegriff.

Definitionen 1:

i) $S_\beta := \{z \in \mathbb{C} \mid |W_\beta(z)| < 1\}$ heißt *Stabilitätsbereich* von β .

ii) Das Verfahren β heißt A-*stabil (absolut stabil)*, falls der Stabilitätsbereich S_β die linke Halbebene enthält, d.h. falls $|W_\beta(z)| < 1$ für alle z aus $\mathbb{C}$ mit $\operatorname{Re} z < 0$ gilt.

A-stabile Verfahren spielen eine wichtige Rolle bei der Behandlung steifer Differentialgleichungssysteme.

2. Spiegelung

Zu $W_\beta(z)$ betrachten wir

(5) $$\tilde{W}_\beta(z) := \{W_\beta(-z)\}^{-1}$$

(Null- bzw. Polstellen sind gesondert zu betrachten) und zu S_β entsprechend

(6) $$\tilde{S}_\beta := \{z \in \mathbb{C} \mid |\tilde{W}_\beta(z)| < 1\} \; .$$

Aus (5) und (6) folgt die Beziehung

(7) $$\tilde{S}_\beta = \{z \in \mathbb{C} \mid -z \notin S_\beta\} \; .$$

Der Übergang von S_β zu $\tilde{S}_\beta$ bedeutet zunächst eine Spiegelung des Randes von S_β $(|W_\beta(z)|=1)$ an der imaginären Achse und dann eine Vertauschung der Bereiche mit $|W_\beta(-z)| < 1$ und $|W_\beta(-z)| > 1$.

Am Beispiel des Euler-Verfahrens β ($y_{m+1}=y_m+hf(y_m)$) wird dies durch folgende Skizze verdeutlicht:

Skizze 2:

Spiegelung des Stabilitätsbereiches des Euler-Verfahrens β .

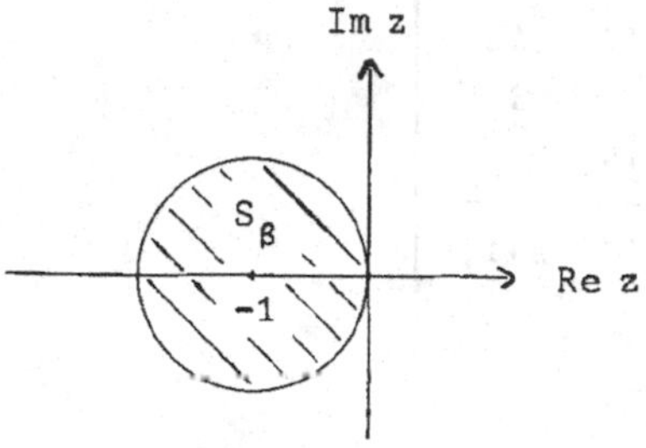

$W_\beta(z)=1+z$

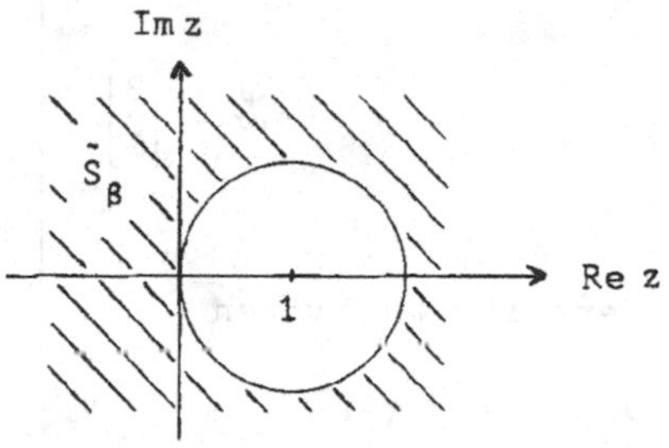

$\tilde{W}_\beta(z)=\frac{1}{1-z}$

Für das rückwärtsgenommene Euler-Verfahren $\tilde{\beta}$ ($y_{m+1}=y_m+hf(y_{m+1})$) gilt $W_{\tilde{\beta}}=\tilde{W}_\beta$, $S_{\tilde{\beta}}=\tilde{S}_\beta$ und $\tilde{\beta}$ ist A-stabil.

Es stellt sich die Frage, ob zu jedem Verfahren β ein Verfahren $\tilde{\beta}$ mit der Eigenschaft $S_{\tilde{\beta}}=\tilde{S}_\beta$ ($W_{\tilde{\beta}}=\tilde{W}_\beta$) existiert? Für Runge-Kutta-Verfahren werden wir eine positive Antwort geben.

Die Untersuchung der Stabilitätsfunktion W_β führt in vielen Fällen auf Padé-Approximationen P_{jk} für die Exponentialfunktion (vgl. CHIPMAN [5], EHLE [7],[8]). Die P_{jk} werden in der *Padé-Tafel* folgendermaßen angeordnet:

$$\left|\begin{matrix} P_{oo} & P_{o1} & P_{o2} & \cdots \\ P_{1o} & P_{11} & P_{12} & \cdots \\ P_{2o} & P_{21} & P_{22} & \cdots \\ \vdots & \vdots & \vdots & \end{matrix}\right| .$$

Die Elemente der Diagonale, der ersten und zweiten Unterdiagonale sind A-*verträglich* (vgl. EHLE [7]), d.h. es gilt $|P_{j+\nu,j}(z)|<1$ für alle z aus $\mathbb{C}$ mit $\mathrm{Re}\, z<o$ ($\nu=o,1,2$; $j=o,1,2,\ldots$). Aus der Betrachtung des Zähler- und Nennerpolynoms von P_{jk} ergibt sich leicht die Beziehung

$$P_{jk}(z)=\{P_{kj}(-z)\}^{-1} . \tag{8}$$

In der Padé-Tafel bedeutet also der Übergang von W_β zu $\tilde{W}_\beta$ (vgl. (5)) eine Spiegelung an der Diagonale, d.h. falls $W_\beta=P_{kj}$ gilt, so folgt $\tilde{W}_\beta=P_{jk}$.

3. Runge-Kutta-Verfahren

Das s-stufige RKV β mit der erzeugenden Matrix

$$(9)\qquad \mathcal{L} = \begin{bmatrix} B \\ \hline G \end{bmatrix} = \begin{bmatrix} b_{11} & \dots & b_{1s} \\ \vdots & & \vdots \\ b_{s1} & \dots & b_{ss} \\ \hline b_{s+1,1} & \dots & b_{s+1,s} \end{bmatrix}$$

liefert die Näherungen

$$(10)\qquad y_{m+1} = y_m + h\, G\,\overline{U}' \ ,$$

wobei

$$(11)\qquad \overline{U}' := \begin{bmatrix} U_1' \\ \vdots \\ U_s' \end{bmatrix} ,\ U_j' := f(U_j)\ ,\ \overline{U} := \begin{bmatrix} U_1 \\ \vdots \\ U_s \end{bmatrix} := \begin{bmatrix} y_m \\ \vdots \\ y_m \end{bmatrix} + h\, B\,\overline{U}'$$

gilt. Die Stabilitätsfunktion W_β des RKV β besitzt die Darstellung (vgl. STETTER [9]S.132)

$$(12)\qquad W_\beta(z) = \frac{\det[I - z\mathcal{L}]^*}{\det[I - zB]} \quad \left([I - z\mathcal{L}]^* := \begin{bmatrix} I - zB & \begin{matrix} 1 \\ \vdots \end{matrix} \\ \hline -zG & 1 \end{bmatrix} \right) .$$

Auf der Menge $\mathcal{R}$ aller RKV werden ein Produkt $\beta_1\beta_2$ (Hintereinanderausführung) und eine Multiplikation $t\,\beta$ $(t \in \mathbb{R})$ definiert (vgl. BUTCHER [3],[4], STETTER [9]S.122). Hier interessieren nur das Verfahren $-\beta$ und das *inverse* Verfahren β^{-1}, welche durch $-\mathcal{L}$ und

$$(13)\qquad \mathcal{L}_{-1} := \begin{bmatrix} B - \begin{bmatrix} G \\ \vdots \\ G \end{bmatrix} \\ -G \end{bmatrix}$$

erzeugt werden. Für die Stabilitätsfunktion gilt

$$(14)\qquad W_{-\beta}(z) = W_\beta(-z)$$

bzw.

$$(15)\qquad W_{\beta^{-1}}(z) = \{W_\beta(z)\}^{-1} \ .$$

Zusammengenommen erhalten wir folgenden

Satz 3:

Zu jedem RKV β mit der erzeugenden Matrix $\mathcal{L}$ existiert das RKV $\bar\beta := -\beta^{-1}$ mit der erzeugenden Matrix

$$(16) \qquad \tilde{\mathcal{L}} = \begin{bmatrix} b_{s+1,s}-b_{ss} & \cdots & b_{s+1,1}-b_{s1} \\ \vdots & & \vdots \\ b_{s+1,s}-b_{1s} & \cdots & b_{s+1,1}-b_{11} \\ b_{s+1,s} & \cdots & b_{s+1,1} \end{bmatrix}$$

so, daß $S_{\bar\beta} = \bar S_\beta$ $(W_{\bar\beta} = \bar W_\beta)$ *gilt.*

Definition 4:

Das RKV $\bar\beta := -\beta^{-1}$ heißt das zu β *gespiegelte* Verfahren.

Die Spiegelung $\Psi: \beta \to \bar\beta$ ist eine eineindeutige Abbildung von $\mathcal{R}$ in $\mathcal{R}$; außerdem gilt $\bar{\bar\beta} = \beta$. Bei symmetrischen RKV $(\bar\beta = \beta)$ besteht die Beziehung $S_{\bar\beta} = S_\beta$ $(W_{\bar\beta} = W_\beta)$; z.B. sind die Gauß-Formeln (vgl. BUTCHER [1]) symmetrisch.

Aus Betrachtungen über die Ordnung p zusammengesetzter RKV folgt

Satz 5:

Die Ordnung des RKV β bleibt bei der Spiegelung Ψ erhalten, d.h. es gilt $p_{\bar\beta} = p_\beta$.

4. Anwendungen

Die Spiegelung Ψ wird vorteilhaft zur Konstruktion A-stabiler RKV verwendet. Außerdem werden Zusammenhänge zwischen verschiedenen Formeln hergestellt. Der Bereich S_β bzw. die Funktion W_β ist oft einfacher zu untersuchen als $S_{\bar\beta}$ bzw. $W_{\bar\beta}$. Als Beispiel seien die impliziten RKV vom Radau- bzw. Lobatto-Typ der Ordnung 2s-1 bzw. 2s-2 angeführt. Die von BUTCHER [2] hergeleiteten Formeln sind durch erzeugende Matrizen mit Nullen in der ersten Zeile (d.h. U_1' ist explizit gegeben) oder in der s-ten Spalte $(b_{s+1,s} \neq 0)$ (d.h. U_s' ist explizit gegeben) bzw. mit Nullen in der ersten Zeile und s-ten Spalte $(b_{s+1,s} \neq 0)$ gekennzeichnet. Unter Berücksichtigung der Ordnung und der Beziehung (12) ergeben sich die zugehörigen Stabilitätsfunktionen sofort als Elemente der ersten und zweiten Oberdiagonale der Padé-Tafel. Mit Hilfe der Spiegelung Ψ erhalten wir stark A-stabile RKV der Ordnung 2s-1 bzw. 2s-2. Diese Verfahren sind mit von EHLE [7] angegebenen identisch, aber dieser neue Zugang ist wesentlich einfacher und eleganter.

Literatur

1. Butcher, J.C.: Implicit Runge-Kutta processes. Math. Comp. 18, 50-64 (1964).
2. Butcher, J.C.: Integration processes based on Radau quadrature formulas. Math. Comp. 18, 233-244 (1964).
3. Butcher, J.C.: The effective order of Runge-Kutta methods. Conf. on the numer. sol. of diff. eqns., Lecture Notes in Mathematics No. 109, 133-139, Springer 1969.
4. Butcher, J.C.: An algebraic theory of integration methods. Math. Comp. 26, 79-106 (1972).
5. Chipman, F.H.: Numerical solution of initial value problems using A-stable Runge-Kutta processes. Thesis, Univ. of Waterloo 1971.
6. Dahlquist, G.: A special stability problem for linear multistep methods. BIT 3, 27-43 (1963).
7. Ehle, B.L.: On Padé approximations to the exponential function and A-stable methods for the numerical solution of initial value problems. Thesis, Univ. of Waterloo 1969.
8. Ehle, B.L.: A-stable methods and Padé approximations to the exponential. SIAM J. Math. Anal. 4, 671-680 (1973).
9. Stetter, H.J.: Analysis of discretization methods for ordinary differential equations. Berlin-Heidelberg-New York: Springer 1973.

Rudolf Scherer
Mathematisches Institut
der Universität Tübingen
Auf der Morgenstelle 10
D-7400 Tübingen 1

ON FAST POISSON SOLVERS AND APPLICATIONS

Johann Schröder, Ulrich Trottenberg, Kristian Witsch

Mathematisches Institut der Universität zu Köln

Weyertal 86-90, D-5000 Köln 41

1. Introduction

Fast Poisson solvers, i.e., fast direct or semi-direct methods to solve problems involving the discrete Poisson equation, recently attracted considerable interest. In particular, so-called *reduction methods* have been developed. As examples we mention Hockney's FACR methods [10], the Block-Cyclic Reduction of Buzbee, Golub and Nielson [5] (also see Golub [9]) and the widely used algorithm of Buneman [5]. (A survey on direct Poisson solvers is given by Dorr in [7].) Although one can apply these methods directly only to rather special problems, such as Dirichlet's problem on a rectangle, the methods, nevertheless, serve as a tool in solving more complicated problems by iterative procedures and capacitance matrix methods.

Independent of the authors mentioned above, our group in Cologne developed a series of reduction methods. It was one of the goals of the meeting on the "Numerical Treatment of Differential Equations" in Oberwolfach 1976 to intensify the contact between mathematicians from various places who are interested in fast Poisson solvers and related subjects.

This paper combines the three talks given by the authors in Oberwolfach and provides a survey on the results which so far have been achieved by our group. This group consists not only of the three authors; we shall try to give appropriate credit to all who contributed to the work. In particular, results from several Diplomarbeiten are incorporated (see the list at the end of the paper). Also, a brief "historical" survey is added below.

The reduction method which we studied most intensively is called *Total Reduction*. Section 2 of this paper describes the basic concept of Total Reduction and other reduction methods, such as *Partial Reduction* and *Alternating Reduction*. Section 3 provides details of our computer programs and numerical results; Section 4 contains statements on the stability of Total Reduction. To simplify the presentation we restrict ourselves in these three sections to Poisson's equation on a square with homogeneous boundary conditions. However, most results hold for more general problems. (For example, other boundary conditions are considered in Section 5.1.)

Section 5 is concerned with a series of generalizations and applications. We treat Helmholtz' equation and the biharmonic equation. Moreover, applications to problems with more general regions and to nonlinear problems are discussed. We describe our approach to these problems and report on numerical results. A large part of this work was carried out in Diplomarbeiten; certain programs and results are not yet optimal. Nevertheless, we found it worthwhile to inform the reader about some of the achievements.

As already mentioned above, Total Reduction is the reduction method that we studied most intensively. Our numerical experience indicates that the method of Alternating Reduction is also very promising. In particular, Alternating Reduction can be applied to three-dimensional problems (see Section 5.6). - Finally,Partial Reduction is equivalent to Block-Cyclic Reduction.

The concept of Total Reduction, based on a calculus of difference stars, was conceived in 1955 by Schröder [13]. At that time, however, only two special cases were published in scientific journals: the reduction method for certain ordinary differential equations [14] and the application of only one total reduction step in the case of Poisson's equation [12] (see Sections 2.1 and 2.4 below). These methods seemed suitable for desk calculators, which were the most advanced means of computation available to the author.

The work on Total Reduction was resumed 1970 in the Diplomarbeit of P. Schmidt [D9]. In writing his Fortran program, Schmidt for the first time used truncated stars (see Section 3.1). In addition, he proved some stability results, in particular, for the phase of solution. A systematic study of Total Reduction and other reduction methods was finally undertaken by Schröder and Trottenberg in [15],[16]. Both these papers were written in close cooperation. Nevertheless, Schröder is mainly responsible for the development of reduction methods and the calculus of difference stars in the first paper [15], while Trottenberg is mainly responsible for the stability theory in the second paper [16]. H. Reutersberg, the third author of [16], wrote fast new programs (TR 1 and TR 2), carried out numerical tests and provided numerical material for the stability theory. He also discovered formula (3.2) for the limit star S'_{∞} . W. Seyfert [D11] wrote programs for Alternating Reduction in two and three dimensions. K. Witsch participated in our teamwork and contributed in many ways to the development of the reduction methods and the applications as described in Section 5 below and in Section 7.2 of [16]. In particular, Witsch supervised several Diplomarbeiten.

2. General description of reduction methods

2.1 Reduction for an ordinary differential equation

Some ideas related to reduction methods are very easily explained for ordinary differential equations. Therefore, we first consider the simple differential equation

$$-\varphi''(x) = f(x) \quad \text{on the interval} \quad (0,1)$$

for functions φ which satisfy the homogeneous Dirichlet boundary conditions $\varphi(0) = \varphi(1) = 0$. - This section is also intended to motivate the notation used in the sections below.

Applying the ordinary difference method or a common Hermite formula to the given problem one obtains a system of equations of the form

$$-u(x_{i-1}) + 2u(x_i) - u(x_{i+1}) = r_0(x_i) \quad (i = 1,2,\ldots,n-1) \tag{2.1}$$

with $x_i = ih$, $h = n^{-1}$, $u(x_0) = u(x_n) = 0$, where, respectively,

$$r_0(x_i) = h^2 f(x_i) \quad \text{or} \quad r_0(x_i) = \frac{h^2}{12}(f(x_{i-1}) + 10f(x_i) + f(x_{i+1})) \; .$$

The unknown function u , defined on the *grid* $\{x_0, x_1, \ldots, x_n\}$, approximates φ .

These difference equations can be written as

$$[-1 \quad 2 \quad -1]u(x_i) = r_0(x_i) \quad \text{or} \quad S_0 u(x_i) = r_0(x_i) \tag{2.2}$$

with a *difference operator* S_0 , determined by the *difference star* $[-1 \quad 2 \quad -1]$ and the given grid.

We consider three succeeding equations

$$\begin{array}{l|lllll}
1 & [-1 \quad 2 \quad -1]u(x_{i-1}) & & & = r_0(x_{i-1}) \\
2 & \quad [-1 \quad 2 \quad -1]u(x_i) & & & = r_0(x_i) \\
1 & \qquad [-1 \quad 2 \quad -1]u(x_{i+1}) & & & = r_0(x_{i+1})
\end{array}$$

and build a linear combination of them using the factors 1,2,1, indicated on the left-hand side. In this way we obtain a new equation

$$[-1 \quad 0 \quad 2 \quad 0 \quad -1]u(x_i) = r_1(x_i) \quad \text{with} \quad r_1(x_i) = [1 \quad 2 \quad 1]r_0(x_i) \; ; \tag{2.3}$$

that means

$$-u(x_{i-2}) + 2u(x_i) - u(x_{i+2}) = r_1(x_i) \quad \text{with} \quad r_1(x_i) = r_0(x_{i-1}) + 2r_0(x_i) + r_0(x_{i+1}) \; .$$

The difference star $[-1 \quad 0 \quad 2 \quad 0 \quad -1]$ can be considered as the *product* $[1 \quad 2 \quad 1][-1 \quad 2 \quad -1]$.

If n is an even number, equations (2.3) taken for x_i with $i=2,4,\ldots,n-2$ constitute a system for the unknowns $u(x_i)$ at these points only. Thus the number of unknowns has been *reduced*. When this *reduced system* is solved, the remaining unknowns are obtained from (2.1).

Now, the reduced system essentially has the same form as the given system, so that a further *reduction step* may be carried out, provided also $^n/_2$ is an even number. If n is a power of 2, $n = 2^p$, one can continue this reduction process until only one equation for the unknown $u(\frac{1}{2})$ remains.

After the *reduction* has been *completed* in this way, one calculates the unknown values of u at $x = \frac{1}{2};\frac{1}{4},\frac{3}{4};\frac{1}{8},\frac{3}{8},\frac{5}{8},\frac{7}{8};\ldots$ one after the other in this order.

This reduction method has been described in [14] for stars of the form $[-1 \quad d \quad -1]$. The method can be interpreted as a Gauß elimination procedure with a specific choice of the pivot elements. This procedure can be generalized to solve difference equations for the more general differential equation $-(p(x)\varphi'(x))' + q(x)\varphi(x) = f(x)$ (see [15]). Then, however, an important feature of the reduction method as described above is lost. For difference equations with *constant difference star* independent of x, the matrix of the system need not actually be transformed. Instead, the reduced equations are obtained using a calculus for difference stars, such as in

$$[1 \quad d \quad 1][-1 \quad d \quad -1] = [-1 \quad 0 \quad d_1 \quad 0 \quad -1] \quad \text{with } d_1 = d^2 - 2 .$$

That the elimination process (on the left-hand side of the system) can be replaced by computing certain difference stars, becomes very important for partial differential equations.

2.2 The difference equations for a Dirichlet-problem

Consider the problem

$$-\Delta\varphi(P) = f(P) \quad \text{on the unit square} \quad \Omega = \{P = (x,y) : 0 < x,y < 1\}$$

with homogeneous Dirichlet boundary conditions. (For inhomogeneous conditions, see Section 5.1.) Let G_0 be the *infinite grid*

$$G_0 = \{(x_i,y_j) : x_i = ih \ , \ y_i = jh \ ; \ i,j = 0, \pm 1, \pm 2, \ldots\}$$

with $h = n^{-1}$, $n = 2^p$, $p \in \mathbb{N}$, and Ω_0 the *finite grid* $\Omega_0 = G_0 \cap \Omega$.

The usual *five-point formula* or the *usual nine-point formula* leads to a system of equations of the form

$$S_0 u(P) = r_0(P) \tag{2.4}$$

for $P = (x,y) \in \Omega_0$ where, respectively,

$$S_0 = \begin{bmatrix} & -1 & \\ -1 & 4 & -1 \\ & -1 & \end{bmatrix} \quad \text{or} \quad S_0 = \begin{bmatrix} -1 & -4 & -1 \\ -4 & 20 & -4 \\ -1 & -4 & -1 \end{bmatrix} \tag{2.5}$$

Here u and r_0 denote Ω_0-*grid-functions*, i.e., real-valued functions defined on Ω_0. The function r_0 is determined by the given function f and the difference method used. The function u is also defined in the grid-points Q on the boundary, where $u(Q) = 0$.

The Ω_0-grid-functions u and r_0 can be extended to G_0-grid-functions u and r_0 (defined on the infinite grid) such that the *difference equations* (2.4) *hold for all* $P \in G_0$. The extended functions are odd functions with respect to each of the variables x and y; they have the period 2 in x and y direction; and they vanish at the points of G_0 which belong to the boundary $\partial\Omega$. In other words, both u and r_0 have the *extension property* described by

$$\left.\begin{aligned} &u(x,y) = -u(x,-y) = -u(-x,y) \,, \\ &u(x+2k, y+2\ell) = u(x,y) \quad \text{for } k,\ell = 0, \pm 1, \pm 2 \,, \\ &u(P) = 0 \quad \text{for } P \in \partial\Omega \,. \end{aligned}\right\} \quad (2.6)$$

The reduction method explained below starts from the system of equations (2.4) on G_0. The process of reduction is best described in terms of a calculus of difference stars and difference operators.

2.3 Difference stars and difference operators

The above quantities S_0 are examples for *difference stars* and, at the same time, for *difference operators*.

A *difference star*

$$S = \begin{bmatrix} & \vdots & \vdots & \vdots & \\ \cdots & s_{-1,1} & s_{01} & s_{11} & \cdots \\ \cdots & s_{-1,0} & s_{00} & s_{10} & \cdots \\ \cdots & s_{-1,-1} & s_{0,-1} & s_{1,-1} & \cdots \\ & \vdots & \vdots & \vdots & \end{bmatrix} = [s_{ij}]$$

is simply a scheme of numbers, the *elements* of S. For given grid G, a difference star describes a *difference operator*, $u(x,y) \to \Sigma\, s_{ij} u(x+ih, y+jh)$, which maps the set of all G_0-grid-functions into itself. For example,

$$\begin{bmatrix} & -1 & \\ -1 & 4 & -1 \\ & -1 & \end{bmatrix} u(P) = 4u(x,y)-u(x-h,y)-u(x+h,y)-u(x,y-h)-u(x,y+h) \,.$$

For a given grid, the letter S will also be used to name the corresponding difference operator, $u \to Su$. To indicate the grid we may also write

$S = [s_{ij}]_{G_0}$ for the difference operator.

The product operator TS defined by $(TS)u = T(Su)$ is described by the *product star*

$$TS = ST = W = [w_{ij}] \quad \text{with} \quad w_{ij} = \sum_{\mu\nu} s_{\mu\nu} t_{i-\mu, j-\nu} \; .$$

To a star $S = [s_{ij}]$ we define the *conjugate star* $\bar{S} = [(-1)^{i+j} s_{ij}]$, which is obtained by changing the signs in a chess-boardlike manner. Also, let $S^+ = \frac{1}{2}(\bar{S} + S)$, $S^- = \frac{1}{2}(\bar{S} - S)$. For a given grid, the *conjugate difference operator*, also named $\bar{S}$, and the difference operators S^+, S^- are defined correspondingly. (We emphasize that for a difference operator the operations $\bar{\ }, {}^+, {}^-$ depend on the grid considered.)

Example.

$$S = \begin{bmatrix} -1 & -4 & -1 \\ -4 & 20 & -4 \\ -1 & -4 & -1 \end{bmatrix}_{G_0} , \quad \bar{S} = \begin{bmatrix} -1 & +4 & -1 \\ +4 & 20 & +4 \\ -1 & +4 & -1 \end{bmatrix}_{G_0} ,$$

$$\bar{S}S = \begin{bmatrix} 1 & 0 & -14 & 0 & 1 \\ 0 & -72 & 0 & -72 & 0 \\ -14 & 0 & 340 & 0 & -14 \\ 0 & -72 & 0 & -72 & 0 \\ 1 & 0 & -14 & 0 & 1 \end{bmatrix}_{G_0} , \quad \text{or} \quad \bar{S}S = \begin{bmatrix} & & 1 & & \\ & -14 & -72 & -14 & \\ 1 & -72 & 340 & -72 & 1 \\ & -14 & -72 & -14 & \\ & & 1 & & \end{bmatrix}_{G_1} . \qquad (2.7)$$

We see that the star which describes the difference operator $\bar{S}S$ on G_0 has zero-elements destributed in a chess-boardlike manner ($\bar{S}S$ is an *even difference operator* on G_0, i.e., $\bar{S}S = (\bar{S}S)^+$). Because of this property, this difference operator also maps the set of all G_1-grid-functions into itself, where G_1 is the grid with mesh-width $h_1 = \sqrt{2}h$ explained in Figure 1. Thus, this difference operator on G_1 can be described by a difference star $[s_{ij}]_{G_1}$ with respect to G_1, as on the right-hand side of (2.7).

For any given difference operator S *on* G_0, $\bar{S}S$ *maps the set of* G_1*-grid functions into itself.* For brevity, we shall say: $\bar{S}S$ is an operator on G_1.

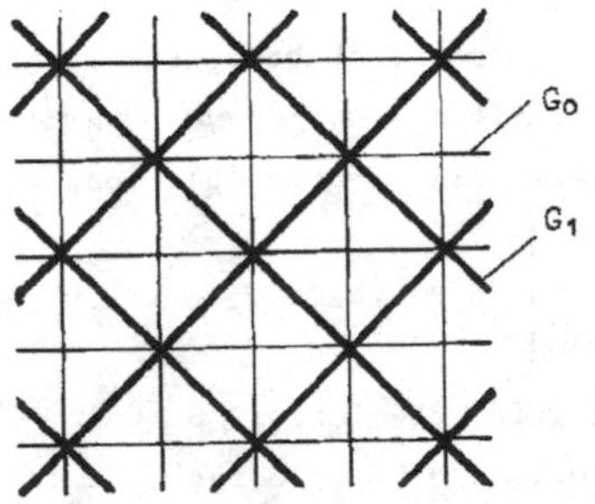

Figure 1:
grids G_0 (lines parallel to the axes) and G_1 (diagonal lines)

2.4 Applying one reduction step

We start with the difference equations

$$S_0u(P) = r_0(P) \quad \text{for} \quad P \in G_0 \tag{2.8}$$

constructed above. By applying the conjugate operator $\bar{S}_0$ to this set of equations (i.e., to the grid functions S_0u and r_0) we obtain a new set of equations. Of these equations we consider only those for $P \in G_1$, that is,

$$S_1u(P) = r_1(P) \quad \text{for} \quad P \in G_1 \quad \text{with} \quad S_1 = \bar{S}_0S_0 \ , \quad r_1 = \bar{S}_0r_0 \ . \tag{2.9}$$

As we have seen above, S_1 maps the set of G_1-grid-functions into itself. Consequently, the latter system only contains unknowns $u(Q)$ with $Q \in G_1$. We have carried out one *reduction step.*

Described in different words we have done the following. We have built certain linear combinations of the given equations (2.8) such that all unknowns $u(Q)$ which do not belong to points $Q \in G_1$ are eliminated. This was possible by choosing the elements $(-1)^{i+j}s_{ij}$ of the conjugate star $\bar{S}_0$ as factors in the said linear combinations.

In general, one will carry out further reduction steps, as explained in the next section. Let us describe, however, in which way the unknowns $u(P)$ $(P \in \Omega_0)$ can be calculated if only one reduction step has been applied.

We first solve the system

$$S_1u(P) = r_1(P) \quad \text{on the \textit{finite} grid} \quad \Omega_1 = G_1 \cap \Omega \tag{2.10}$$

to obtain the values $u(Q)$ for $Q \in \Omega_1$, and then solve the system

$$S_0u(P) = r_0(0) \quad \text{on the \textit{remaining set}} \quad R_0 = \Omega_0 \sim \Omega_1$$

to obtain the values $u(Q)$ with $Q \in R_0$. The latter system can also be written as

$$S_0^+u(P) = S_0^-u(P) + r_0(P) \quad \text{for} \quad P \in R_0 \ . \tag{2.11}$$

The term $S_1u(P)$ in (2.10) evaluated at a point $P \in \Omega_1$ may involve values $u(Q)$ for points $Q \in G_1$ which do not belong to Ω_1 . However, because of the extension property (2.6), this value $u(Q)$ equals the value of u or $-u$ at some point of Ω_1 . Thus, if these relations (2.6) are exploited, *the system* (2.10) *becomes a set of equations for the unknowns* $u(Q)$ *with* $Q \in \Omega_1$ *only*. Compared with the given system (2.4) on Ω_0 , *the number of unknowns has been reduced* approximately by a factor $^1/_2$.

When the *reduced system* (2.10) has been solved, the terms $S_0^-u(P)$ in the *remaining* system (2.11) are known, since here only values $u(Q)$ with $Q \in G_1$ are involved. The term $S_0^+u(P)$ on the left-hand side contains only values $u(Q)$ with $Q \in R_0 = \Omega_0 \sim \Omega_1$, if again the extension property (2.6) is exploited.

We see that the matrix of the linear system obtained from (2.10) and (2.11) has the form

$$\begin{array}{|c|c|} \hline A_1 & 0 \\ \hline K_0 & H_0 \\ \hline \end{array} \quad \text{with square sub-matrices } H_0, A_1 \,. \tag{2.12}$$

Observe that, to carry out only this one reduction step, the extension property (2.6) has not been used to its full extent. Indeed, with S_0 describing the five-point formula, only points $Q \in G_0$ occurred which either belong to $\bar{\Omega}$ or have at most the distance h from $\bar{\Omega}$. For that reason, the method explained above can also be applied to five-point difference equations for *more general domains* Ω , as described in [12] .

Further points $Q \notin \bar{\Omega}$ will be used, however, when several reduction steps are done.

2.5 The reduction completed

When the system (2.9) on G_1 is obtained from the system (2.8) on G_0 , no use is made of the way the G_0-grid-functions u and r_0 have been constructed. In particular, the given domain Ω and the extension property (2.6) are not involved. For this step of the procedure, u and r_0 may be any functions defined on G_0 .

Therefore, we may now consider the system (2.9) on G_1 to be the given system and apply a further reduction step to this system, obtaining in this way

$$S_2u(P) = r_2(P) \quad \text{for } P \in G_2 \text{ , with } S_2 = \bar{S}_1S_1 \text{ , } r_2 = \bar{S}_1r_1 \text{ .}$$

Here G_2 is related to G_1 , as G_1 is related to G_0 ; more precisely, $G_2 = \{(x_i,y_j) : i,j = 0, \pm 2, \pm 4,\dots\}$. Of course, now all terms such as $\bar{S}_1$ etc. are to be understood with respect to G_1 . For example, if $S_1 = \bar{S}S$ in (2.7), then

$$\bar{S}_1 = \begin{bmatrix} & & 1 & & \\ & -14 & 72 & -14 & \\ 1 & 72 & 340 & 72 & 1 \\ & -14 & 72 & -14 & \\ & & 1 & & \end{bmatrix}_{G_1} ,$$

This procedure may be continued so that equations

$$S_ku(P) = r_k(P) \quad \text{for } P \in G_k \text{ , } k = 0,1,\dots,\mu \text{ with } \mu \le m \text{ ,} \tag{2.13}$$

$m = 2(p-1)$, are obtained where

$$S_k = \bar{S}_{k-1} S_{k-1} \ , \quad r_k = \bar{S}_{k-1} r_k \tag{2.14}$$

and $\bar{S}_k$ is the conjugate operator with respect to G_k . The points $P \in G_k$ are distributed in G_{k-1} in a chess-boardlike manner.

The question arises whether these systems (2.13) can be used to obtain a system for the unknowns $u(P)$ $(P \in \Omega_0)$ which can be solved even more easily than the system described by the matrix (2.12). It turns out that this is possible due to the special structure of the grid $(n = 2^p)$ and the extension property of the functions. The ideas involved are the same as for the first reduction step. For example, applying the second reduction step results in transforming the (square) sub-matrix A_1 of (2.12) in the same way as the matrix A_0 of the given system (2.4) was transformed in the first reduction step.

After μ reduction steps one obtains a system with a block-triangular matrix

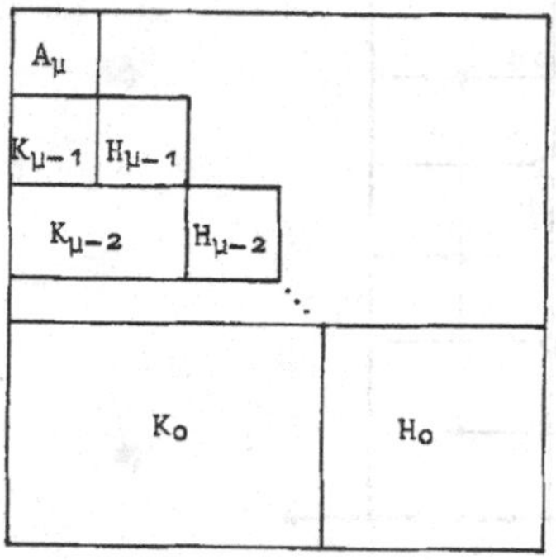

with square matrices A_μ and H_k . (2.15)

For $\mu = 1$, this is the matrix in (2.12). For $\mu = m$, the matrix A_m obtained last consists of only one element. *For* $\mu = m$, *the reduction is completed.*

The system for the unknowns $u(P)$ $(P \in \Omega_0)$ which is obtained after m reductions can be written in a form analogous to (2.10),(2.11):

$$S_m u(P) = r_m(P) \qquad \text{for} \quad P \in \Omega_m \tag{2.16}$$

$$S_k^+ u(P) = S_k^- u(P) + r_k(P) \ , \quad \text{for} \quad P \in R_k = \Omega_k \sim \Omega_{k+1} \tag{2.17}$$

where $k = m-1, m-2, \ldots, 1, 0$.

For example, let $n = 8$ and $m = 4$. Then G_4 has the mesh-width $\frac{1}{2}$, and $\Omega_4 = G_4 \cap \Omega$ contains only the point marked by ◆ in Figure 2. The value u at this midpoint is calculated by solving the single equation (2.16). Of course, here the extension property (2.6) has to be used; the term $S_4 u(P)$ contains values in further points Q which, however, all either lie on the boundary $\partial\Omega$ or have the form $Q = (\frac{1}{2} + \ell , \frac{1}{2} + q)$ with integers ℓ, q .

G_3 has the mesh-width $\frac{1}{4}\sqrt{2}$, and $\Omega_3 = G_3 \cap \Omega$ consists of the points marked by ◆ or o , so that $R_3 = \Omega_3 \sim \Omega_4$ consists of the points marked by o . The values $u(P)$ for $P \in R_3$ are obtained by solving (2.17) for $k = m - 1 = 3$.

Next, the values in R_2 (marked by □) are obtained, after that the values in R_1 (marked by •), and finally the values in R_0 (not marked specifically). Observe that in case of the five-point formula the set (2.17) with $k = 0$, solved last, consists of single equations each of which contains only one unknown.

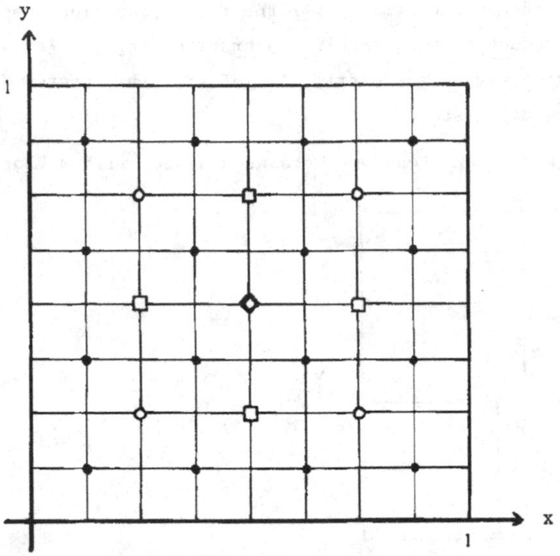

Figure 2

In most of our numerical examples, we used $n \leq 128$ (see Sections 3.3 and 5). For $n = 128$ one needs $m = 12$ reduction steps.

2.6 Total reduction, TR-method

The reduction method described in the sections above is called *Total Reduction*, in contrast to *Partial Reduction* and other reduction methods explained below. If completed, the method of Total Reduction will simply be called *TR-method* here. Several versions of this method have been implemented (see Section 3.2). Our programs have the structure now explained.

The stars $S_k, \bar{S}_k, S_k^+, S_k^-$ neither depend on the given function f , nor on the mesh-width h . Thus, they can be stored to be available when a concrete problem is solved.

Solving a concrete problem then essentially consists of two parts:

1. *Phase of reduction:* The grid functions $r_k(P)$ on Ω_k are calculated using the second formula in (2.14).

2. *Phase of solution:* The system of difference equations (2.16),(2.17) with the block-triangular matrix (2.15) is solved. Since all diagonal blocks *are highly diagonally dominant*, each sub-system is solved by an iterative method.

The phase of reduction can be viewed as an elimination procedure which transforms the given system (2.4) into a system with the block-triangular matrix (2.15). The important point is that by this process the given system, which has a *large condition number* $\approx$ const h^{-2} , is transformed into a *"harmless"* system, and that this is done in a *numerically stable* way (see Section 4).

2.7 Generalizing the concept of reduction

In the method of Total Reduction the number of unknowns is reduced in each step by applying a conjugate difference star. Hence, the *operation of conjugation* which transforms a star S into its conjugate star $\bar{S}$ is an important concept of our theory. We shall now investigate this operation a little further and consider similar operations in order to generalize the whole approach.

One sees immediately that the transformation of $S=[s_{ij}]$ into $\bar{S}=[(-1)^{i+j}s_{ij}]$ can be carried out in two steps. One may first change the signs of the elements in each second line in x-direction to obtain the star $[(-1)^i s_{ij}]$ and afterwards change the signs in each second line in y-direction to obtain $[(-1)^j((-1)^i s_{ij})]=\bar{S}$. In other words, one applies two operations C_x and C_y , defined by

$$C_xS = [(-1)^i s_{ij}] , \quad C_yS = [(-1)^j s_{ij}] .$$

We call C_x the operation of *x-conjugation* and C_xS the *x-conjugate star* to S and use a corresponding terminology for C_y and C_yS . Furthermore,

$$C_{(x,y)} = C_xC_y = C_yC_x$$

is the *(x,y)-conjugation* and $C_{(x,y)}S$ the *(x,y)-conjugate star* (which here coincides with $\bar{S}$).

Example:

$$S = \begin{bmatrix} -1 & -4 & -1 \\ -4 & 20 & -4 \\ -1 & -4 & -1 \end{bmatrix}, \quad C_xS = \begin{bmatrix} 1 & -4 & 1 \\ 4 & 20 & 4 \\ 1 & -4 & 1 \end{bmatrix}, \quad C_yS = \begin{bmatrix} 1 & 4 & 1 \\ -4 & 20 & -4 \\ 1 & 4 & 1 \end{bmatrix}. \tag{2.18}$$

So far, we considered only (grid-)functions of the two variables (x,y) . In this situation it is natural to say that C_xS is a *partial conjugate star* (the *x-partial conjugate star*), whereas $C_{(x,y)}S$ is the *total conjugate star*. For a given grid G_0, each star corresponds to a difference operator as explained in Section 2.3. Therefore, we shall also use terms such as *x-partial conjugate difference operator* (with respect to G_0), etc.

Now, these partial conjugate stars or difference operators can also be used to reduce the number of unknowns. This is done by exploiting the following important Property (R).

(R) *The multiplication of a given difference operator (on a grid* G_0 *) by any of its partial or total conjugates yields a difference operator on a coarser subgrid.*

As we have already seen in Section 2.3, $(C_{(x,y)}S) \cdot S = \bar{S}S$ represents an operator on G_1 with G_1 explained in Figure 1.

On the other hand, the star $(C_xS)S$ has zero-elements in each second column, so that it also represents a difference operator on the grid $\tilde{G}_1 = \{(x_i, y_j) : i = 0, \pm 2, \pm 4, \ldots ; j = 0, \pm 1, \ldots\}$ with mesh-width $2h$ in x-direction and mesh-width h in y-direction.

Example: For S in (2.18), we have

$$(C_xS) \cdot S = \begin{bmatrix} -1 & 0 & 14 & 0 & -1 \\ -8 & 0 & -176 & 0 & -8 \\ -18 & 0 & 396 & 0 & -18 \\ -8 & 0 & -176 & 0 & -8 \\ -1 & 0 & 14 & 0 & -1 \end{bmatrix}_{G_0} = \begin{bmatrix} -1 & 14 & -1 \\ -8 & -176 & -8 \\ -18 & 396 & -18 \\ -8 & -176 & -8 \\ -1 & 14 & -1 \end{bmatrix}_{\tilde{G}_1} \qquad (2.19)$$

Because of Property (R) the reduction method can be generalized in a way described as follows. (Here, the terms $S_k, G_k, \ldots$ $(k = 1,2,\ldots)$ will be redefined; they need not have the same meaning as in the previous sections.)

A general reduction step: Given a system (2.4)

$$S_0 u(P) = r_0(P) \qquad (2.20)$$

of difference equations on a grid G_0, choose any (partial or total) conjugate $\tilde{S}_0$ of S_0 and apply $\tilde{S}_0$ to the given equations to obtain

$$S_1 u(P) = r_1(P) \qquad (2.21)$$

with $S_1 = \tilde{S}_0 S_0$, $r_1 = \tilde{S}_0 r_0$. If S_1 is an operator on a sub-grid G_1, the latter equations taken for $P \in G_1$ build a system for the unknowns $u(Q)$ $(Q \in G_1)$ only. One may solve this system and afterwards calculate the unknowns $u(Q)$ for $Q \in G_0 - G_1$ by solving the original equations (2.20) on $G_0 - G_1$. The type of such a reduction step depends on the conjugate $\tilde{S}_0$ chosen. We shall use the notation *total reduction step, partial reduction step,* C_x*-reduction step* or *x-reduction step,* etc. in an obvious way.

All that we have explained so far remains meaningful if G_0 is replaced by a parallelogram-grid of congruent parallelograms consisting of two families of parallel lines, two of which are distinguished as the (x,y)-coordinate system. For that reason the process of reduction may be continued as follows.

One may apply a further reduction step to the system (2.21) on G_1, choosing any conjugate $\tilde{S}_1$ of S_1; afterwards the system so obtained may again be reduced, etc. The operators $\tilde{S}_0, \tilde{S}_1, \ldots$ used in this process are (partial or total) conjugates with respect to the corresponding grids $G_0, G_1 \ldots$. These conjugates need not be of the same kind; one may use partial and total conjugates in an arbitrary sequence. We thus have a *series of reduction methods*.

Of course, when the reduced equations (2.21) are solved, one has to exploit the extension property (2.6) in order to obtain a finite system for the unknowns $u(Q)$ with $Q \in \Omega_1 = G_1 \cap \Omega$. For that purpose, the domain Ω has to "fit" into the subgrid G_1. The same holds for the grids $G_2, G_3, \ldots$. In each case it suffices to choose G_0 with $n = 2^p$ as in Section 2.2. For certain procedures, however, this restriction on G_0 may be relaxed considerably.

The general reduction step differs formally from a step of total reduction only by the fact that an arbitrary conjugate $\tilde{S}_0$ is used instead of $\bar{S}_0$. However, the reduced system (2.21) may be of a quite different nature depending on the choice of $\tilde{S}_0$. Consequently, the class of reduction methods contains procedures which are quite different from each other.

Three reduction methods are of particular interest:

1. The method of *Total Reduction* considered above, where only total reduction steps are applied.

2. The method of *Partial Reduction* where only partial reduction steps of the same kind are applied; that means either only x-reductions or only y-reductions.

3. The method of *Alternating Reduction* where x- and y-reduction steps are applied alternatingly.

2.8 Partial Reduction

Without loss of generality, we consider only Partial Reduction in x-direction (*x-Partial Reduction*). Also, we assume that S_0 is one of the stars in (2.5), although much of the formalism remains valid for arbitrary S_0. The x-Partial Reduction is described by formulas (2.13),(2.14),(2.16),(2.17) when the following changes are made. Define now

$$G_k = \{(x,y) : x = i2^k h \ ; \ \ y = jh \ ; \ \ i,j = 0, \pm 1, \pm 2, \ldots\} \ , \quad m = p-1$$

and replace $\bar{S}_k$ by $\tilde{S}_k = C_x S_k$, where the conjugation C_x is defined with respect to G_k. In particular, now let $S_k^+ = \frac{1}{2}(\tilde{S}_k + S_k)$, $S_k^- = \frac{1}{2}(\tilde{S}_k - S_k)$.

For simplicity, we restrict ourselves to the case $\mu = m$, for which the Partial Reduction is completed (*PR-method*). The system then obtained can again be described by a matrix of type (2.15).

These similarities to the TR-method, however, are only formal ones; there are essential differences from a numerical point of view. First, *the unknowns are calculated in a different order;* and second, most of *the diagonal blocks* of the final matrix (2.15) here *have a very bad condition* $\approx \exp n$. In particular, these submatrices are far from being diagonally dominant. For example, for $S_0 = S$ in (2.18) we have

$$S_1^+ = \begin{bmatrix} 14 \\ -176 \\ 396 \\ -176 \\ 14 \end{bmatrix}$$

and the elements of this star are used to build the elements in the rows of H_1.

Partial Reduction, in principle, is not restricted to "small" stars S_0 such as those in (2.5). For these stars, however, the algorithm becomes simpler. For example, the unknowns can be calculated line-wise. First, the unknowns which belong to $x = \frac{1}{2}$ are obtained, afterwards those for $x = \frac{1}{4}$, $x = \frac{3}{4}$, etc., the sequence $x = \frac{1}{2};\frac{1}{4},\frac{3}{4};\frac{1}{8},\frac{3}{8},\frac{5}{8},\frac{7}{8};\ldots$ being the same as for the reduction method for ordinary differential equations (Section 2.1). Also, for stars as in (2.5) we need not extend the grid-functions in x-direction.

For these reasons, the organization of the PR-method is simpler than that of the TR-method. On the other hand, the instability of the PR-method is a serious disadvantage (see [16], Sections 5.4 and 6.2).

Since no reduction in y-direction is involved, x-Partial Reduction can be applied in the same way to rectangles of side length $2^p h$ in x-direction and νh in y-direction, where the integer ν need not be a power of 2.

Partial Reduction essentially is equivalent to the method of *(Block-)Cyclic Reduction* or *(Cyclic) Odd-Even-Reduction* described by Buzbee, Golub and Nielson [5] (Section 3) in matrix terminology (also see Golub [9]). These authors also recognized the stability problems involved. The *CORF algorithm* [5] and the *Buneman algorithm* [2],[5] also are closely related to Partial Reduction. (See Section 2.11 concerning factorization.) However, Buneman's algorithm, which is now widely used, has better stability properties than Partial Reduction itself (see the examples in [16], Section 6.2). While the paper of Buzbee, Golub and Nielson clarifies the theoretical background of the methods considered, Buneman's algorithm can be considered as a stabilization of the CORF algorithm.

2.9 Alternating Reduction

The method of Alternating Reduction (AR-method), where x- and y-reductions alternate, also can be described by formulas similar to (2.13) through (2.17). The details can be worked out without difficulty. The important question is:what are the numerical properties of this method?

Obviously, AR-Reduction has features in common with both TR-Reduction and PR-Reduction. It shares with TR-Reduction the good stability properties (see Section 4). Moreover, the stars $S_0, S_2, S_4, \ldots$ for AR-Reduction are exactly the same as for TR-Reduction. On the other hand, only partial reduction steps are carried out, so that no grids are needed whose axes build angles of 45° with the given coordinate axes. This results in a simpler organization of the algorithm. So far only test programs have been applied.

A definite advantage of AR-Reduction, however, is that it extends to problems of higher dimension.

2.10 Three-dimensional Problems

Many of the concepts in the previous sections carry over to problems of higher dimension, in particular, three-dimensional problems. However, there are differences too. For example, there is no immediate generalization of TR-Reduction to three dimensions.

In the previous sections we made use of the fact that, for any of the reduction methods, all grids G_k which occur in the process are essentially of the same type as G_0. This is not so in three dimensions. The points of a given cubic grid G_0

can be partitioned into two classes such that two neighbor points always belong to different classes, as in spacial chess. Therefore, one total reduction step can be carried out in essentially the same way as for two dimensions, so that one obtains a set of equations for the unknowns which belong to one of the point-classes called G_1. This "grid" G_1 is an octaedric-grid and thus of a totally different structure. Therefore, one cannot continue in the same way.

Of course, one may continue differently. Observe, for example, that reductions can also be carried out for triangular grids which have a structure totally different from that of a square grid (see [15], Section 6.3).

Moreover, there are a series of possibilities to combine partial reduction steps to a reduction method. For example, one may apply alternatingly (x,y)-reduction and z-reduction. All these possibilities are presently being investigated by H. Reutersberg [11].

One of the most promising methods seems to be the *Alternating Reduction* where x-, y- and z-reduction steps are repeated in this order. This method has already been programmed and applied to concrete problems (see Section 5.6).

2.11 Reduction without extending the grid-functions

If a reduction method is implemented, the extension property (2.6) of the grid-functions has to be taken into account by appropriate means. For example, a modulo-function can be used or an index table. The latter possibility was chosen for our algorithms TR 1 and TR 2, which were first written in Fortran suitable for IBM 370/168 (see Section 3.2). Another way is to use the binary representation of the point-indices i,j together with mask- und shift operations in order to determine whether $u(x_i,x_j)$ equals $u(P)$ or $-u(P)$ at a point $P \in \bar{\Omega}$ and what this point P is. This way is being used by Reutersberg in his CDC Cyber program for three-dimensional problems.

The question arises, however, whether the explicit use of this extension cannot be avoided at all. It turns out that this goal can be achieved by properly splitting the difference stars into smaller stars.

For example, the stars S_k^+ which occur in x-Partial Reduction for the five-point star in (2.5) can be written as a product of the form

$$S_k^+ = \begin{bmatrix}-1\\ a_{k1}\\ -1\end{bmatrix} \begin{bmatrix}-1\\ a_{k2}\\ -1\end{bmatrix} \cdots\cdots \begin{bmatrix}-1\\ a_{k,2^k}\\ -1\end{bmatrix},$$

so that the solution of the system (2.17) can be calculated by solving 2^k tridiagonal systems, without any extension being necessary. This method is analogous to a factorization method described in [5] in matrix terminology (CORF algorithm).

Factorizations of this type are also possible for more general stars (see [15], Section 8.1).

However, for TR-Reduction and other reduction methods, another possibility seems more promising. Each two-dimensional star S can be written as a polynomial of the basic stars

$$X = [1 \quad 0 \quad 1] \ , \quad Y = \begin{bmatrix} 1 \\ 0 \\ 1 \end{bmatrix} .$$

For example, if S_0 is the five-point star in (2.5), we obtain for Total Reduction

$$S_0 = 4 - X - Y \ , \quad \bar{S}_0 = 4 + X + Y \ , \quad S_1 = 16 - X^2 - 2XY - Y^2$$

(see [15], Section 8.3). Obviously, if X or Y is applied at a point of Ω_0, no function evaluation at a point outside $\bar{\Omega}$ is necessary. This makes it plausible that by this representation the explicit use of the extension can be avoided. In a somewhat modified form, this polynomial representation has already been incorporated into our TR-algorithms.

3. Programmed TR-algorithms and numerical results

In this section we shall describe implementations of the TR-method for the difference equations (2.4) with the five-point star S_0 in (2.5). These algorithms rely on special properties of the corresponding difference stars S_k .

Similar programs can be written for the nine-point star. Our program TR/M for this star has essentially the same structure and numerical properties as the program TR 2 described below. (Also see [16], Section 7.1.)

3.1 Numerical properties of the difference stars S_k

The number of non-vanishing elements of the difference stars S_k rapidly increases with k (roughly equal to 2^k). Therefore, at first sight it seems difficult to use these stars in an algorithm. However, the stars S_k show a certain convergence behavior which makes them very suitable for numerical treatment.

The normalized stars $S_k' = [n_{ij}^{(k)}] = [s_{ij}^{(k)}/s_{oo}^{(k)}]$ *converge to a "limit" star* S_∞' ,

$$|S_\infty' - S_k'| \to 0 \quad (k \to \infty),$$

with respect to the "star norm"

$$|S| = \sum_{i,j \in \mathbb{Z}} |s_{ij}| \; . \tag{3.1}$$

Here $S_\infty' = [n_{ij}]$ *is given by*

$$n_{ij} = \zeta_{ij} \exp\left(-\tfrac{\pi}{2}(i^2 + j^2)\right) \quad \text{with} \quad \zeta_{ij} = \begin{cases} 1 & \text{if } i \text{ and } j \text{ even} \\ -1 & \text{otherwise.} \end{cases} \tag{3.2}$$

(This convergence statement was recently proved by H. Zimare.) For example, $|S_\infty' - S_{10}'| \approx 10^{-3}$, $|S_\infty' - S_{20}'| \approx 10^{-6}$.

This behavior of the stars S_k has two important consequences.

1. One can use *truncated stars* $S_k(\sigma), \bar{S}_k(\sigma), S_k^+(\sigma), S_k^-(\sigma)$ instead of $S_k, \bar{S}_k, S_k^+, S_k^-$. Here, $S(\sigma)$ is obtained from $S = [s_{ij}]$ when all elements s_{ij} with $|s_{ij}| \le \sigma |s_{oo}|$ are replaced by 0 .

For example, if $\sigma = 10^{-5}$, all truncated stars $S_k(\sigma)$ have no more than 25 non-vanishing elements; all their elements for which $i > 2$ or $j > 2$ equal zero. The relative errors $\tau_k(\sigma) = |S_k - S_k(\sigma)|/|S_k|$ have (at most) the same magnitude as σ:

$$\tau_k \le 2 \cdot 10^{-5} \; , \quad \tau_k \to \tau_\infty \approx 0.9 \cdot 10^{-5} \quad \text{if} \quad \sigma = 10^{-5} \; . \tag{3.3}$$

2. For *each star* S_k^+ the coefficient $s_{oo}^{(k)}$ dominates the sum of the absolute values of the other elements by a factor $q_k \ge 3$, where $q_k \to q_\infty \approx 5.5$. For this reason, the diagonal blocks H_k in (2.15) are highly diagonally dominant, in a way

essentially independent of n .

It should be noted that the stars S_k which belong to the nine-point star S_0 in (2.5) have properties similar to those described above. In particular, the corresponding normalized stars S_k' converge even faster to the same limit-star S_∞' .

3.2 TR-algorithms

Besides an older version TR 1, two Fortran programs TR 2 and TR 3 are used now. Some characteristic details are described below.

In all the programs scaled approximations for the stars S_k $(k = 0,1,2,\dots,m)$ are stored. The accuracy of the approximations insures that the programs can read the truncated stars as accurately as needed in the calculations. The scaling is necessary to prevent the stored elements from becoming too large.

The main operation in both the reduction and solution phase consists in *applying a star to a grid function.* The programs mainly differ in the way in which this operation "star • grid function" is carried out. Further differences concern the iterative methods involved, the special treatment of the first stars $S_0,\dots,S_3$ and the admissible parameters, as will now be described for TR 2 and TR 3.

<u>TR 2.</u> For all stars $\bar{S}_k, S_k^+, S_k^-$ with index $k \geq 4$ the operation star • grid function is here performed in a straightforward manner. The extension of the grid functions outside $\bar{\Omega}$ is simulated by appropriate index tables. For the stars with index $k \leq 3$ a polynomial representation is used, as outlined in Section 2.11. The truncation parameter σ is used as an input parameter.

The subsystems (2.17) are solved by a SOR method with parameters ω_k and iteration numbers ℓ_k which are independent of n . The iterations start with 0 as initial approximation. The numbers ℓ_k either are input data or quantities computed according to the desired iteration accuracy ε . For example,

$$\ell_1 = 1 \; , \quad \ell_2 = 5 \; , \quad \ell_k = 4 \quad \text{for} \quad k \geq 3 \; , \quad \text{if} \quad \varepsilon = 10^{-5} \; . \tag{3.4}$$

The standard version of TR 2 uses $\varepsilon = \sigma$. (Compare the numerical results in the next section.)

<u>TR 3.</u> In this program, worked out by H.F. Dressler, all stars are truncated with a fixed $\sigma = 0.7 \cdot 10^{-4}$, and *a polynomial representation is used for these truncated stars.*

All versions of the TR-method mentioned above need $\vartheta(n) = O(n^2)$ *arithmetic operations for the reduction and solution phase.*

For example, let us consider the reduction phase. For the truncated stars $\bar{S}_k(\sigma)$ the number of non-zero elements is bounded independently of k. Since $\bar{S}_{k-1}(\sigma)$ is applied on Ω_k and Ω_k has at most $n^2 2^{-k}$ points, $O(n^2)$ operations are needed.

Similar arguments hold for the solution phase, since the iteration numbers ℓ_k are independent of n and uniformly bounded in k.

More precisely, one obtains

$$\vartheta(n) \leq 32n^2 \quad \text{for TR 2} \quad (\sigma = \varepsilon = 10^{-5}) \,, \quad \vartheta(n) \leq 30n^2 \quad \text{for TR 3.}$$

For decreasing σ or ε the number of operations increases like $-\log \sigma$ or $-\log \varepsilon$, respectively.

The *storage requirement* is essentially $2n^2$ words.

3.3 Numerical results

The numerical results of this section demonstrate the characteristic properties of the TR-algorithm. The results concern the following simple but typical examples:

(1) $-\Delta\varphi = x(1-x) + y(1-y)$ (on Ω), $\varphi = 0$ (on $\partial\Omega$), with solution $\varphi = x(1-x)y(1-y)/2$;

(2) $-\Delta\varphi = 0$ (on Ω), $\varphi = 1$ (on $\partial\Omega$), with solution $\varphi = 1$.

There is no discretization error for either example so that $u^* = \varphi$ also solves the discrete problem. (The assumptions of Section 2.2 are not fulfilled in Example (2) as the boundary conditions are inhomogeneous; Section 5.1 shows how to treat this case by simple modifications of r_0.)

Table 1 demonstrates the influence of the truncation parameter σ, the iteration accuracy ε and the rounding errors (characterized by the computing accuracy eps). We give the relative errors

$$d_n = \|u_n^* - u_n\| / \|u_n^*\| \tag{3.5}$$

where u_n^* is the exact solution of the discrete problem (on Ω_0), u_n is the approximation computed by TR 2, and $\| \; \|$ denotes the maximum norm on Ω_0. The specification "ε, eps small" means that these parameters are chosen so small that they have no influence on the accuracy of the solution compared with the parameter σ; the other headings have an analogous meaning. In particular, $\varepsilon = 10^{-5}$ represents the iteration numbers in (3.4). The computations for the last column were done in single precision on the IBM 370/168 at the KFA, Jülich; all other computations were carried out on the CDC Cyber 76 at the University of Cologne. (The programs TR 2 and TR 3, which originally were written for an IBM computer, were adapted to the CDC Cyber 76 by G. Lickfeld.)

Table 1

	d_n					
	$\sigma = 10^{-5}$ ε, eps small		$\varepsilon = 10^{-5}$ σ, eps small		eps $\approx 0.5 \cdot 10^{-6}$ σ,ε small	
n	(1)	(2)	(1)	(2)	(1)	(2)
16	.22 (-6)	.50 (-6)	.14 (-5)	.29 (-5)		.92 (-5)
32	.14 (-5)	.32 (-5)	.52 (-6)	.21 (-5)		.10 (-4)
64	.26 (-5)	.32 (-5)	.14 (-5)	.22 (-5)		.17 (-4)
128	.40 (-5)	.99 (-5)	.17 (-5)	.25 (-5)		.19 (-4)

The results show that *the errors have the same magnitude as* σ *or* ε if one of these parameters prevails; *the influence of the rounding errors increases very slowly with* n .

Table 2 contains relative errors d_n and the computing time t_n for TR 2 (with $\varepsilon = \sigma = 10^{-5}$) and TR 3. (The time for generating r_0 is not included.) *Use of the polynomial representation results in a remarkable increase in speed.*

Table 2 (CDC Cyber 76)

n	TR 2 ($\sigma = \varepsilon = 10^{-5}$)			TR 3 ($\sigma = 0.7 \cdot 10^{-4}$, $\varepsilon = 10^{-5}$)		
	d_n		t_n [sec]	d_n		t_n [sec]
	(1)	(2)		(1)	(2)	
16	.15 (-5)	.31 (-5)	.020	.26 (-4)	.29 (-4)	.013
32	.14 (-5)	.32 (-5)	.065	.12 (-4)	.15 (-4)	.042
64	.26 (-5)	.32 (-5)	.265	.20 (-4)	.38 (-4)	.162
128	.52 (-5)	.12 (-4)	1.077	.33 (-4)	.46 (-4)	.639

Other examples which we calculated gave similar results. For these examples the relative errors were even smaller than in Example (2) (see [15], Table 7 and [16], Table 2.3). Several results of the TR-method have been compared with those of other efficient direct (and iterative) methods in [16], Section 6.2. It was found that TR 3 needs approximately the same computing time as the widely used Buneman algorithm [5].

4. Numerical stability of the TR-method

Investigating the numerical stability by theoretical means, we restrict ourselves to the TR-method with the *five-point* star S_0. Similar results also hold for the nine-point star S_0 and the corresponding AR-method.

For our estimates, we use maximum norms $\|v\|$ for functions v, indicating the grid or domain if necessary; for example, $\|v\|^{\Omega_k} := \max\{|v(P)| : P \in \Omega_k\}$. Operators defined on the respective (grid-) function spaces are described by matrices as in (2.15), and the norms of these operators are given by matrix norms. (These matrix norms do not depend on the special enumeration of the grid-points.) We emphasize, however, that all operator norms which occur depend on n. For some estimates involving $\bar{S}_k$ we use the star norm (3.1) which is, of course, independent of n.

4.1 Some general remarks on condition and numerical stability

The condition number (with respect to the maximum norm) of the given system is

$$\alpha_0 = \alpha_0(n) = \|A_0^{-1}\| \, \|A_0\| = O(n^2), \text{ where } \alpha_0(n)/n^2 \nearrow c_0 . \qquad (4.1)$$

Here $c_0 \approx 0.589$ is the maximum value of the solution z of $-\Delta z = 8$ (on Ω), $z = 0$ (on $\partial\Omega$). Using this condition number α_0 one can estimate the influence of arbitrary perturbations of the matrix elements and the right-hand side r_0. In the case under discussion the matrix elements are given exactly, so that

$$\|\delta u\| / \|u\| \leq \alpha_0(n) \|\delta r_0\| / \|r_0\| .$$

This estimate is realistic, if $r_0(x,y) = \gamma_0 \sin(n-1)\pi x \sin(n-1)\pi y$, $\delta r_0(x,y) = \gamma_1 \sin \pi x \sin \pi y$ (with some constants γ_0, γ_1). However, for a boundary value problem with given f, the right-hand side $r_0 = h^2 f$ cannot have the above form if the mesh-width h is chosen reasonably small.

For *fixed* f, an equation $\|\varphi\|^{\Omega} = k_f \|f\|^{\Omega}$ holds. Using this equation (and the convergence $u \to \varphi$ as $h \to 0$) instead of $\|u\| \geq \|r_0\| / \|A_0\|$ one obtains

$$\|\delta u\| / \|u\| \leq \alpha_f \|\delta r_0\| / \|r_0\| \quad \textit{with } \alpha_f \textit{ independent of } n$$

instead of the above estimate. This means that the condition of the discrete problem with respect to the true input data f is bounded independently of n.

For this reason a numerical method is conceivable where the numerical errors do not increase as fast as the condition number $\alpha_0(n) \approx \text{const} \cdot n^2$. (This possibility was also pointed out by Babuska [1].)

We shall now discuss the corresponding behavior of the TR-method.

Arguments similiar to those above apply to problems with inhomogeneous boundary conditions.

4.2 Errors in the TR-method, stability of the phase of solution

When the TR-method is applied, we have to take into account the following sources of errors.

Due to the truncation of the conjugate stars $\bar{S}_k$ and to rounding errors in the phase of reduction, one obtains perturbed grid functions $\tilde{r}_k$ instead of r_k. (In general, one also starts with some perturbed $\tilde{r}_0$.) We denote by $\tilde{u}$ the solution of the reduced system (2.16),(2.17) with r_m, r_k replaced by $\tilde{r}_m, \tilde{r}_k$. Then $\delta_R u := u - \tilde{u}$ is the *error of the phase of reduction.*

Solving the reduced system, one obtains an approximation $\tilde{u} - \delta_S u$ for $\tilde{u}$. The *error of the phase of solution* $\delta_S u$ is caused by truncating the stars S_m, S_k^+ and S_k^-, stopping the iteration and the rounding errors.

The phase of solution is numerically "harmless". This was proved in [16] (Section 4) by detailed estimates for $\delta_S u$. The essential quantities which describe the behavior of $\delta_S u$ are the condition numbers of the diagonal blocks A_m, H_k, the numbers

$$\zeta_k := \max(1, \|H_k^{-1}\| \; \|K_k\|) \tag{4.2}$$

(see (2.15)), and the norms of the iteration matrices belonging to H_k. The condition numbers of all diagonal blocks turn out to be ≤ 2, all numbers $\zeta_k \leq 1.0\bar{6}$, and all Gauß-Seidel iteration matrices have norms ≤ 0.2, with most of these norms much smaller (see [16] for details).

4.3 The operation star • grid function

The phase of reduction, which is the crucial part of the TR-method, consists in successively applying the operation star • grid function. As shown in [16] (Section 3.1), the errors in this operation can easily be estimated such that the estimates hold for any reasonable implementation (such as TR 2 and TR 3).

For an arbitrary star S let T denote the star obtained by rounding the elements of $S(\sigma)$. Moreover, define $w = Sv$ and let $\tilde{w}$ be the function resulting from computing Tv numerically. Then

$$\|w - \tilde{w}\| \leq \eta |S| \; \|v\|$$

with $\eta = \eta(S, \sigma, eps)$. (See [16] for details.)

Applying this estimate to $S = \bar{S}_{k-1}$ and $v = \tilde{r}_{k-1}$, we obtain

$$\|\delta r_k\|^{\Omega_k} \leq \eta_k |\bar{S}_{k-1}| \; \|\tilde{r}_{k-1}\|^{\Omega_{k-1}} \text{ with } \delta r_k = \bar{S}_{k-1}\tilde{r}_{k-1} - \tilde{r}_k . \tag{4.3}$$

The values η_k occurring above differ from the numbers τ_k in (3.3) by a factor which reflects the rounding errors. For the CDC program with $\sigma = 10^{-5}$ this factor is approximately 1 ; for the IBM program see [16], Section 3.1.

4.4 Stability of the phase of reduction

In [16], Sections 3.3 and 3.4, we derived an estimate for the relative error $\|\delta_R u\| / \|u\|$ of the phase of reduction without any assumptions on the form of r_0. The resulting bound behaves roughly like the condition number $O(n^2)$ of the given system.

On the other hand, we obtained bounds of the order $o(n)$ for certain "model problems" where $r_0 = h^2 f$ for a fixed eigenfunction f of $-\Delta$. The estimates for these model problems can be extended to the general case $r_0 = h^2 f$ where f is arbitrary.

Let $\eta_k \leq \eta \quad (k = 1,2,\ldots,m)$, *and* $\|\delta r_0\| \leq \eta_0 \|r_0\|$. *Then*

$$\|\delta_R u\|^{\Omega_0} \leq \frac{c_0}{8} \left[\eta_o + \eta(1+\eta)^{m-1}(1+\eta_0)\mu(n)n^{\omega}\right] \|f\|^{\Omega_0} \tag{4.4}$$

with $c_0 \approx 0.589$ *in* (4.1),

$$\omega \approx 0.595$$

and

$$\mu(16) \leq 2.42, \quad \mu(32) \leq 2.77, \quad \mu(64) \leq 3.00, \quad \mu(128) \leq 3.16, \quad \mu(256) \leq 3.25 .$$

Explicitely,

$$\omega = 2 \log_2 \lambda , \quad \lambda = \frac{|S|}{4 - 2|S^+|} = \frac{\vartheta_3^2(o, e^{-\pi/2})}{4 - 2(\vartheta_3^2(o, e^{-\pi/2}) - \vartheta_2^2(o, e^{-\pi}))} \approx 1.229$$

with $S = S'_\infty$ *in* (3.2) *and Jacobi's theta functions* ϑ_2 *and* ϑ_3 .

We conjecture that the numbers $\mu(n)$ are bounded by $\mu' = 1/(\lambda^2 - \lambda) \approx 3.56$. This hypothesis, however, has not yet been proved.

We shall not prove the above estimate in detail here, but indicate how to modify the arguments in [16] to obtain it.

By induction one obtains from (4.3)

$$\|\delta r_k\|^{\Omega_k} \leq \left(\eta_k \prod_{j=o}^{k-1} (1+\eta_j)|\bar{S}_j|\right) \|r_0\|^{\Omega_0} .$$

This inequality, together with $r_0 = h^2 f$, is used instead of formulas (3.11),(3.12) in [16]. Methods similar to those in [16] then yield

$$\|\delta_R u\|^{\Omega_0} \leq \frac{c_0}{8} \left[\eta_0 + \sum_{k=1}^{m} \eta_k \prod_{j=o}^{k-1} (1+\eta_j)\zeta_j q_j\right] \|f\|^{\Omega_0} \tag{4.5}$$

with ζ_j in (4.2) and $q_j = \|A_{j+1}^{-1}\| \cdot |\bar{S}_j| / \|A_j^{-1}\|$. The behavior of $\zeta_j(n)$ and $q_j(n)$ has been described in [16], Section 3.4.3. For example, $\zeta_j(n) \cdot q_j(n) \leq 1.6$ (if $n \leq 256$) . However, to derive (4.4) from (4.5) and to obtain the explicit representation of ω we used some asymptotic relations involving S_∞' . (Also compare formulas (3.29) in [16], where $\bar{\zeta}'b' = \lambda$.)

The results in [16], Section 3.6, concerning the model problems are obtained from (4.5) by using $\|f\|^{\Omega_0} \doteq (\nu_1^2 + \nu_2^2)\pi^2 \|\varphi\|^{\Omega_0}$, $\|\varphi\|^{\Omega_0} \leq \|u\|^{\Omega_0}$.

5. Modifications and applications

Throughout this section S_g will denote the five-point star in (2.5), to which we have restricted ourselves for simplicity.

5.1 Other boundary conditions

In Section 2 we considered only homogeneous Dirichlet boundary conditions. We now explain how to proceed, when inhomogeneous Dirichlet or periodic boundary conditions are given.

In the case of an *inhomogeneous Dirichlet boundary condition* $\varphi = g$ on $\partial\Omega$ we again obtain the equations (2.4) on Ω_0 (with $S_0 = S_g$, $r_0 = h^2 f$). However, now $u = g$ on $\partial\Omega \cap G_0$. Instead of this system we solve

$$S_0\tilde{u}(P) = \tilde{r}_0(P) \quad (P \in \Omega_0), \quad \tilde{u}(P) = 0 \quad (P \in \partial\Omega \cap G_0)$$

with $\tilde{r}_0(P) = r_0(P) - S_0\tilde{g}(P)$ $(P \in \Omega_0)$, $\tilde{g} := g$ on $\partial\Omega \cap G_0$, $\tilde{g} := 0$ on Ω_0. This system has the same form as assumed in Section 2 and the solution $\tilde{u}$ coincides with u on Ω_0. Although the structure of the right-hand side differs here, the numerical results in [16] and Table 1 show that the TR-method in this case has approximately the same accuracy as for homogeneous boundary conditions.

For *periodic boundary conditions* one can use the TR-Method with some slight modifications (which were programmed by T.G. Wulff for the IBM version of TR 2). Here the boundary value problem can be regarded as a problem on the entire plane where the solution φ and the right-hand side f are extended periodically. By discretization, one immediately obtains a system of difference equations for all points of G_0 with periodically extended grid functions.

The TR-method, now with $m = 2p - 1$ reduction steps, is applied to this singular system. The system is solvable, if the (discrete) mean value of f is zero. After prescribing the value of the solution in one of the corners, one can solve the reduced equations. The remaining algorithmic details are the same as for Dirichlet boundary conditions and the numerical results show approximately the same accuracy. For instance, we obtained the following relative errors d_n (3.5) for the boundary value problem with solution $\varphi = u = 1$:

n	16	32	64	128
d_n	.15 (-4)	.12 (-4)	.12 (-4)	.16 (-4)

These numbers are to be compared with those in the last column of Table 1.

We would like to point out that the periodic boundary value problems play an important role for the capacitance matrix method as developed by Widlund and Proskurowski [18] (see Section 5.4).

5.2 The Helmholtz equation

The Helmholtz equation $-\Delta\varphi + c\varphi = f$ (with a constant c) leads to a system of difference equations (2.4) with $S_0 = S_g + ch^2$. This system can again be handled by Total Reduction. The corresponding stars S_k, however, now depend on c and h. There is an essential difference between problems with $c > 0$ and problems with $c < 0$.

For $c > 0$, the TR-method can be applied in exactly the same manner as in the case $c = 0$. (Special programs for the rapid calculation of the stars S_k were worked out by B. Holzberg [D3].) In particular, the same iteration numbers ℓ_k (3.4) can be used, since the diagonal dominance of the matrices H_k is even higher than for $c = 0$. For instance, for the boundary value problem with solution $\varphi = u = 1$ we obtained by using the IBM version of TR 2

$$d_{64} = 0.11 \cdot 10^{-4} \text{ when } c = 100 \text{ , and } d_{64} = 0.17 \cdot 10^{-4} \text{ when } c = 0$$

(see the last column of Table 1).

If $c < 0$, several problems arise. The difference equations may not be solvable. Also, if c decreases, the possibility to truncate the stars becomes more restricted; and the diagonal dominance of the matrices H_k becomes less distinct or is lost completely. On the other hand, if $c > -\lambda_1$ (with λ_1 the first eigenvalue), the same truncation is possible as for $c = 0$ and the subsystem can be solved by iterative methods.

5.3 Biharmonic equations

We now consider the boundary value problem

$$\text{(P1)} \qquad \Delta\Delta\varphi = f \text{ on } \Omega \text{ ; } \quad \varphi = g_0 \text{ , } \quad -\Delta\varphi = g_1 \text{ on } \partial\Omega$$

and

$$\text{(P2)} \qquad \Delta\Delta\varphi = f \text{ on } \Omega \text{ ; } \quad \varphi = g_0 \text{ , } \quad \frac{\partial\varphi}{\partial\nu} = g_2 \text{ on } \partial\Omega$$

(with the outward normal ν).

5.3.1 Problem (P1). If $-\Delta$ is replaced by the five-point star S_g one obtains a system of difference equations on Ω_0. This system can be extended to

$$S_0 u = r_0 \quad \text{on} \quad G_0 \text{ , with } S_0 = S_g^2 = \begin{bmatrix} & & 1 & & \\ & 2 & -8 & 2 & \\ 1 & -8 & 20 & -8 & 1 \\ & 2 & -8 & 2 & \\ & & 1 & & \end{bmatrix} \qquad (5.1)$$

Here r_0 and u are defined outside $\bar{\Omega}$ as in (2.6). (For inhomogeneous boundary values g_0 and g_1 , r_0 is modified in a way similar to that described in Section 5.1.) We used two methods to solve this system.

The first method consists in a direct application of the TR-method. H. Förster [D1] calculated several examples with an algorithm analogous to the TR 2 program. The results showed the same stability behavior as for TR 2. Moreover, this algorithm also needs $O(n^2)$ arithmetic operations.

The second possibility is to split (5.1) into $S_g v = r_0$, $S_g u = v$ and to solve these two systems by the TR-method. The corresponding results (obtained by using TR 2) were as accurate as for the above algorithm and the computing time was slightly less. However, it should be noted that the first method can also be applied to the equation $\Delta\Delta\varphi + c\varphi = f$, for any constant $c > 0$.

Trottenberg [17] developed a method for splitting certain two-point boundary value problems of the fourth order (with non-constant coefficients) into second order problems, which then are treated by a reduction method.

5.3.2 Problem (P2). Here a direct application of Total Reduction seems to be impossible. However, the TR-method (and other fast direct Poisson solvers) can be used as an auxiliary means in *capacitance matrix methods*. In the method which we used one tries to find a discrete function g_1 so that the solution u of the discrete problem corresponding to (P1) fulfils $B_\nu u(P) = g_2(P)$. This means that u is also the solution of the discrete problem corresponding to (P2) . (We point out that $B_\nu u(P) = g_2(P)$ cannot be required for all $P \in G_0 \cap \partial\Omega$; the four corners and one neighborhood-point for each corner have to be omitted, otherwise the system would be singular.)

The matrix of the linear system which determines the values of g_1 is called the capacitance matrix. This matrix describes the relationship between the given and the auxiliary discrete problem. The matrix elements, which are independent of f, g_0 and g_2 , can be calculated by applying B_ν to certain solutions of the discrete version of (P1) (with $f = 0$, $g_0 = 0$ and several g_1). This calculation (and an LU-factorization of the capacitance matrix) has only to be carried out once in order to solve several problems of the form (P2) .

The method described above is similar to the capacitance matrix method proposed by Buzbee and Dorr [3]. The general concept of capacitance matrices was developed by Buzbee, Dorr, George and Golub in [4], where, however, only applications to the Poisson equation on irregular regions were considered. (See the following section.)

Our method, combined with TR 2, was numerically investigated by M. Schnell [D10] and H. Förster [D1]. The following five approximations B_ν of the normal derivative were used (here written with respect to a point $P = (x,y)$ on the left boundary):

(N 1) $B_\nu u(x,y) = (u(x,y) - u(x+h,y))/h$

(N 2) $B_\nu u(x,y) = (3u(x,y) - 4u(x+h,y) + u(x+2h,y))/2h$

(N 3) $B_\nu u(x,y) = (u(x-h,y) - u(x+h,y))/2h$

(N 4) $B_\nu u(x,y) = (11u(x,y) - 18u(x+h,y) + 9u(x+2h,y) - 2u(x+3h,y))/6h$

(N 5) $B_\nu u(x,y) = (u(x-\tilde{h}/2,y) - u(x+\tilde{h}/2,y))/\tilde{h}\ ,\quad \tilde{h} = 1/(n-1)\ ,$

$\frac{1}{2}(u(x-\tilde{h}/2,y) + u(x+\tilde{h}/2,y)) = g_0(x,y)\ .$

(Notice that the last approximation uses a modified grid with no points on the boundary.) The order of consistency is $O(h), O(h^2), O(h^2), O(h^3)$ and $O(h^2)$, respectively.

<u>Table 3</u>

		(E1)		(E2)		(E3)
B_ν	n	δ_n	ρ_n	δ_n	ρ_n	δ_n
(N 1)	4	.23 (-1)		.17 (-1)		.14 (-13)
	8	.67 (-2)	.29	.60 (-2)	.35	.50 (-13)
	16	.18 (-2)	.27	.26 (-2)	.43	.10 (-12)
	32	.76 (-3)	.42	.13 (-2)	.49	.45 (-12)
(N 2)	4	.20 (-2)		.18 (-2)		.28 (-13)
	8	.46 (-3)	.23	.31 (-3)	.17	.64 (-13)
	16	.11 (-3)	.24	.72 (-4)	.23	.14 (-12)
	32	.28 (-4)	.25	.19 (-4)	.26	.54 (-12)
(N 3)	4	.52 (-3)		.84 (-2)		.71 (-14)
	8	.20 (-3)	.39	.21 (-2)	.25	.36 (-13)
	16	.55 (-4)	.27	.56 (-3)	.27	.43 (-13)
	32	.14 (-4)	.25	.14 (-3)	.25	.40 (-12)
(N 4)	4	.31 (-3)		.34 (-2)		.57 (-13)
	8	.44 (-4)	.14	.21 (-4)	.06	.85 (-13)
	16	.59 (-5)	.14	.30 (-5)	.14	.14 (-12)
	32	.77 (-6)	.13	.41 (-6)	.14	.63 (-12)
(N 5)	4	.36 (-2)		.72 (-2)		.14 (-13)
	8	.21 (-2)	.24	.20 (-2)	.28	.50 (-13)
	16	.51 (-3)	.24	.51 (-3)	.26	.10 (-12)
	32	.12 (-3)	.24	.13 (-3)	.25	.45 (-12)

Table 3 shows some typical numerical results. It gives the relative errors

$$\delta_n := \|\varphi - u_n\|^{\Omega_0} / \|\varphi\|^{\Omega_0}$$

for the boundary value problems with the solutions

(E1) $\varphi(x,y) = e^x \sin y$

(E2) $\varphi(x,y) = e^x + 2y + x^2y^3$

(E3) $\varphi(x,y) \equiv 1$

and the five approximations (N1)-(N5) of the normal derivative.

The examples (E1) and (E2) demonstrate the influence of the discretization errors. For second order convergence, one would expect the ratio $\rho_n = \delta_{2n}/\delta_n$ to be approximately 0.25 (see (N2), (N3), (N5)). The results for the third order approximation (N4) are much more accurate than those of the other approximations. This indicates that it may be advisable to use a third order approximation of the normal derivative even though the approximation of $\Delta\Delta\varphi$ is only of the second order.

Example (E3) contains no discretization error. Thus the given numbers $\delta_n = d_n$ (3.5) demonstrate the numerical accuracy of our method when combined with TR 2 (CDC version).

5.4 Other regions

In the previous sections we assumed the region Ω to be the unit square and considered only square grids. The method of Total Reduction, however, is also applicable if Ω and the grid G_0 are rectangular. The number m of reduction steps which can be carried out depends only on n_x and n_y, the number of intervals in the x- and y-direction, respectively. In each case, the final matrix A_m has a much smaller condition number than A_0. If n_x is either 2^p or $3 \cdot 2^p$ and n_y either 2^q or $3 \cdot 2^q$, then A_m is an (1,1)-matrix.

For more general (irregular) regions Ω, fast Poisson solvers (such as the TR-method) can again be used as an auxiliary means in *capacitance matrix methods*. For example, the region Ω may be *imbedded* into a rectangular region R. An auxiliary boundary value problem on R is then solved instead of the given boundary value problem on Ω. Here again the relationship between these problems is given by the capacitance matrix.

The method of imbedding is described in detail by Buzbee, Dorr, George and Golub in [4], where the linear system associated with the capacitance matrix is solved directly by an LU-factorization. A preliminary version of this direct method was presented by George [8]. Moreover, George [8] investigated an iterative method where the capacitance matrix need not be calculated explicitely.

Some numerical results for these methods were obtained by U. Nickel [D7] using the IBM program TR 1. Table 4 contains results for two L-shaped domains Ω_1 and Ω_2 where squares of side-length $^1/_4$ and $^1/_2$, respectively, have been removed from the upper right corner of the unit square R .

Table 4

		Ω_1		Ω_2	
	n	32	64	32	64
direct:	d_n	.43 (-5)	.10 (-4)	.76 (-5)	.80 (-5)
	t_n [sec]	4.69	31.89	8.95	62.14
	t_n^o [sec]	.58	2.15	.65	2.45
iterative:	d_n	.11 (-4)	.74 (-5)	.16 (-4)	.83 (-5)
	t_n [sec]	1.33	4.59	1.56	5.75
	t_n^o [sec]	1.32	4.55	1.54	5.71

Again d_n denotes the relative error (3.5) for the boundary value problem with solution $\varphi = u = 1$. Moreover, t_n is the total computing time, and t_n^o the total computing time diminished by the time needed to calculate the capacitance matrix and its LU-factorization. For the iterative method, Nickel used the Davidon algorithm with iteration numbers 3 for Ω_1 and 4 for Ω_2 (for both $n = 32$ and $n = 64$).

One sees that the iterative method is much faster, if only one problem has to be solved. The direct method may be prefered if t_n^o is the crucial quantity. This is the case, if the Poisson equation has to be solved several times for the same region Ω , e.g. in an "outer" iteration (see Section 5.5).

Finally, we would like to mention that a modified (and improved) imbedding method was described by Widlund and Proskurowski in [18]. Here the capacitance matrix is calculated by using a Poisson solver for periodic boundary value problems (or the Fourier-Toeplitz method).

5.5 Nonlinear equations

Consider a nonlinear equation

$$-\Delta\varphi(P) = f(P,\varphi(P)) \quad (P \in \Omega)$$

(together with boundary conditions). Under suitable conditions this problem may be solved by using the Picard iteration

$$-\Delta\varphi_{n+1}(P) = f(P,\varphi_n(P)) \quad (P \in \Omega)$$

or a generalized Picard iteration

$$-\Delta\varphi_{n+1}(P) + c\varphi_{n+1}(P) = c\varphi_n(P) + f(P,\varphi_n(P)) \quad (P \in \Omega)$$

(for constant c) or Newton-like methods (with $c = c_n$). In all these processes fast Poisson solvers can be used as auxiliary means for the corresponding discrete equation (if necessary, combined with some capacitance matrix method).

For such purposes, the TR-method is particularly advantageous, because the truncation parameter σ and the iteration parameter ε can be chosen according to the rate of convergence of the "outer" iteration. For instance, if one calculates the corrections $\varphi_{n+1} - \varphi_n$ (more precisely, their discrete approximations), "inner" iteration numbers $\ell_k = 1$ or $\ell_k = 2$ $(k \geq 1)$ are sufficient. This result in a 30 - 40 % saving in computing time for TR 3 and a somewhat smaller saving for TR 2 (see Table 5).

Table 5: Computing time t_n (CDC Cyber 76)

	TR 2, $\sigma = 10^{-5}$		TR 3	
	t_{64} [sec]	t_{128} [sec]	t_{64} [sec]	t_{128} [sec]
$\ell_k = 1$	.178	.704	.103	.395
$\ell_k = 2$	.205	.822	.119	.468
ℓ_k (3.4)	.259	1.050	.160	.637

We point out that iterative procedures combined with fast Poisson solvers can be used to solve much more general (linear and nonlinear) boundary value problems (see, for instance, [6]).

5.6 AR-Reduction

Finally, we report on our first experiences with the methods of Alternating Reduction in two and three dimensions. The respective programs (for the ordinary difference method) were written for the IBM 370/168 by W. Seyfert [D11].

5.6.1 Two-dimensional AR-Reduction. For the AR-Reduction different grids are used than for the TR-Reduction. Also the corresponding stars have less symmetries and slightly different numerical properties than the TR-stars. In particular, the diagonal dominance of the matrices $H_2, H_4, H_6, \ldots$ is less distinct; thus the iteration numbers $\ell_2, \ell_4, \ell_6, \ldots$ have to be chosen approximately 1.5 times larger than for the TR-method to obtain the same accuracy. The matrix H_0 is tridiagonal and the corresponding system therefore is solved directly.

Apart from these differences, the program for the AR-method has a structure similar to that of our TR-programs. For the boundary value problem with solution $\varphi = u = 1$ we obtained the results

$$d_{16} = 0.13 \cdot 10^{-4} , \quad d_{32} = 0.31 \cdot 10^{-4} , \quad d_{64} = 0.44 \cdot 10^{-4}.$$

(These numbers are to be compared with those in the last column of Table 1.) The computing time was slightly longer than for a comparable version of TR 2.

5.6.2 Three-dimensional AR-Reduction. Our program for the three-dimensional AR-method is a straightforward extension of the two-dimensional version. The numerical properties of the corresponding difference stars have not yet been investigated in detail. Therefore, the iteration numbers are not prescribed a priori. Instead, the iterations are stopped when successive approximations $v^{(i)}$ satisfy $\|v^{(i)} - v^{(i-1)}\| / \|v^{(i)}\| \leq 10^{-4}$.

For the three-dimensional boundary value problem with solutions $\varphi = u = 1$ we obtained

$$d_8 = 0.43 \cdot 10^{-4} , \quad d_{16} = 0.10 \cdot 10^{-3} , \quad d_{32} = 0.13 \cdot 10^{-3} .$$

The corresponding iteration numbers $\ell_0, \ell_1, \ell_2; \ell_3, \ell_4, \ell_5; \ell_6, \ldots; \ell_m$ were:

14, 13, 8;	13, 13, 8;	8, 7, 4;	1		for $n = 16$,
14, 13, 8;	14, 14, 8;	11, 10, 6;	8, 7, 4;	1	for $n = 32$.

The computing time was proportional to n^3 , the approximate number of unknowns.

References

[1] BABUSKA, I.: Numerical stability in problems of linear algebra. SIAM J. Numer. Anal. 9, 53-77 (1972).

[2] BUNEMAN, O.: A compact non-iterative Poisson solver. SUIPR Report Nr. 294, Institute for Plasma Research, Stanford University, Stanford, Calif. 1969.

[3] BUZBEE, B.L., DORR, F.W.: The direct solution of the biharmonic equation on rectangular regions and the Poisson equation on irregular regions. SIAM J. Numer. Anal. 11, 753-763 (1974).

[4] BUZBEE, B.L., DORR, F.W., GEORGE, J.A., GOLUB, G.H.: The direct solution of the discrete Poisson equation on irregular regions. SIAM J. Numer. Anal. 8, 722-736 (1971).

[5] BUZBEE, B.L., GOLUB, G.H., NIELSON, C.W.: On direct methods for solving Poisson's equations. SIAM J. Numer. Anal. 7, 627-656 (1970).

[6] CONCUS, P., GOLUB, G.H.: Use of fast direct methods for the efficient numerical solution of nonseparable elliptic equations. SIAM J. Numer. Anal. 10, 1103-1120 (1973).

[7] DORR, F.W.: The direct solution of the discrete Poisson equation on a rectangle. SIAM Rev. 12, 248-263 (1970).

[8] GEORGE, J.A.: The use of direct methods for the solution of the discrete Poisson equation on nonrectangular regions. Rep. STAN-CS-70-159, Computer Science Dept., Stanford University, Stanford, Calif. 1970.

[9] GOLUB, G.H.: Direct methods for solving elliptic difference equations. In: Symposium on the Theory of Numerical Analysis, 1-19. (Lecture Notes in Mathematics 193). Berlin-Heidelberg-New York: Springer 1971.

[10] HOCKNEY, R.W.: The potential calculation and some applications. Methods in Computational Physics 9, 135-211 (1970).

[11] REUTERSBERG, H.: Reduktionsverfahren zur Lösung elliptischer Differenzengleichungen im $\mathbb{R}^3$. Forthcoming.

[12] SCHRÖDER, J.: Zur Lösung von Potentialaufgaben mit Hilfe des Differenzenverfahrens. ZAMM 34, 241-253 (1954).

[13] SCHRÖDER, J.: Beiträge zum Differenzenverfahren bei Randwertaufgaben. Habilitationsschrift Hannover 1955.

[14] SCHRÖDER, J.: Über das Differenzenverfahren bei nichtlinearen Randwertaufgaben I. ZAMM 36, 319-331 (1956).

[15] SCHRÖDER, J., TROTTENBERG, U.: Reduktionsverfahren für Differenzengleichungen bei Randwertaufgaben I. Numer. Math. 22, 37-68 (1973).

[16] SCHRÖDER, J., TROTTENBERG, U., REUTERSBERG, H.: Reduktionsverfahren für Differenzengleichungen bei Randwertaufgaben II. Numer. Math. 26, 429-459 (1976).

[17] TROTTENBERG, U.: Lösung linearer gewöhnlicher Randwertaufgaben vierter Ordnung mit Hilfe von Aufspaltungen. Numer. Math. 25, 297-306 (1976).

[18] WIDLUND, O., PROSKUROWSKI, W.: On the numerical solution of Helmholtz's equation by the capacitance matrix method. ERDA Rep. COO-3077-99, Courant Institute of Mathematical Sciences, New York Univ., 1975.

Diplomarbeiten at the University of Cologne (no copies available)

[D1] FÖRSTER, H.: Direkte Anwendungen von Reduktionsverfahren zur Lösung von Randwertaufgaben vierter Ordnung. 1976.

[D2] HINZEN, K.: Untersuchung und numerische Erprobung des Buneman-Algorithmus. 1974.

[D3] HOLZBERG, B.: Reduktionsverfahren zur Lösung von Differenzengleichungen für die Differentialgleichung $-\Delta u + cu = f(x)$ und ihre Anwendung zur iterativen Lösung von Differenzengleichungen für $-\Delta u = F(x,u)$. 1974.

[D4] KAHL, H.: Lösung gewöhnlicher Randwertaufgaben vierter Ordnung mit Reduktionsverfahren. 1975.

[D5] LICKFELD, G.: Untersuchung verschiedener Verfahren zur Lösung elliptischer Differenzengleichungen. 1976.

[D6] MUCKE, F.U.: Zur CORF-Methode, Beschreibung und numerische Experimente. 1971.

[D7] NICKEL, U.: Anwendung des Reduktionsverfahrens zur Lösung der Poisson-Gleichung auf nicht-rechteckigen Gebieten nach George u.a. 1974.

[D8] REUTERSBERG, H.: Untersuchung und Programmierung eines Reduktionsverfahrens zur Lösung von Differenzengleichungen bei elliptischen Differentialgleichungen. 1973.

[D9] SCHMIDT, P.: Reduktionsmethode und Alternierendes Verfahren zur Lösung von Differenzengleichungen beim Dirichlet-Problem. 1970.

[D10] SCHNELL, M.: Numerische Lösung der Differentialgleichung $\Delta\Delta u = f$ mit Dirichletschen Randbedingungen (Verwendung des Reduktionsverfahrens). 1975.

[D11] SEYFERT, W.: Verfahren der alternierenden partiellen Reduktion. 1975.

CONSIDERATIONS CONCERNING A THEORY FOR ODE-SOLVERS

HANS J. STETTER

TECHNICAL UNIVERSITY OF VIENNA

Abstract: In today's general purpose software packages for initial value problems in ODE's, the course of the computation is normally determined by a *tolerance parameter* δ: The particular integration procedure and the stepsize to be used in the next step are derived from δ and the local behaviour of the ODE. This control mechanism should imply (for a sufficiently wellbehaved ODE) that the global error ε satisfies $\varepsilon(t) = v(t)\delta + o(\delta)$ where v depends on the problem and the package but not on δ. (Naturally round-off is not considered as $\delta \to 0$.) To achieve this "tolerance-convergence", the control procedure has to guarantee that in each step we obtain the exact solution to a δ-perturbed ODE; this might be called "tolerance-consistency". Furthermore, no situation must arise in which the steplengths decrease in a geometric progression. Test computations have established the proportionality requested by tolerance-convergence in a satisfactory manner for the Shampine-Gordon package.

1. Introduction

The classic theory for the numerical solution of ordinary initial value problems by discretization methods uses the *stepsize* h as its central parameter; other quantities are assessed in terms of powers of h as $h \to 0$, which leads to the concept of *order*. Variable steps can be accomodated in this theory through the use of coherent grids (cf. [1]); but still the grids are conceptionally fixed a priori. Furthermore, a *fixed method* is considered.

In constrast to that line of thought, today's software packages for the treatment of initial value problems for systems of ODE's*) typically implement one (or even several) *class(es) of integration procedures*; the selection of the procedure and of the stepsize to be used at a given point in the computation is done a posteriori on the basis of comparisons of locally computed quantities with *prescribed control quantities*. Thus, stepsize and order retain only a local meaning and become dependent rather than independent quantities. An immediate application of the classical theory to an analysis of the global behaviour of such ODE-packages is hence impossible.

The first large-scale analysis of an ODE-package was carried out in the remarkable book [2] by Shampine and Gordon; it indicates the necessity for the formulation of appropriate concepts to account for this

*) We exclusively consider *general purpose* ODE-solvers which have only quantitative problem data available and which pass through the integration interval only once.

fundamental change in attitude. In the following, I will try to discuss various aspects of such a theory for ODE-solvers and suggest some preliminary concepts.

Throughout, we will - for simplicity - assume that the ODE-system is

$$(0) \qquad y' = f(t,y), \quad t \in [0,T], \quad y(0) = y_0 ,$$

and that the true solution y exists and is sufficiently smooth (at least piecewise). Furthermore, f is to be (piecewise) differentiable also in a sufficiently large neighbourhood of the solution trajectory. (Other simplifying assumptions of this kind may not always be explicitly formulated in this preliminary analysis.)

2. The tolerance parameter

What do we expect from a general purpose ODE-package? It should be a "black box" which, when fed the problem data and the abscissas at which approximate values of the solution are requested, delivers these values.

When we consider a situation where the specified interval of integration [0,T] for a given ODE is kept fixed and one further output abscissa t moves from 0 towards T (but does not approach T too closely) then we may assume that the output value $\eta(t)$ of a given ODE-package, for fixed control parameters, is a continuous and piecewise differentiable function of t (if we neglect round-off errors; see end of this section). Therefore we may also consider the global error $\varepsilon(t) := y(t) - \eta(t)$ as a continuous and piecewise differentiable function under these circumstances. If, in some equation involving $\eta(t)$ or $\varepsilon(t)$, t takes the value of a jump point of the first derivative - which will normally happen at a gridpoint of the computation - then either left-hand or right-hand limits have to be taken.

According to the rules of classical numerical analysis, the package should also deliver bounds, or at least estimates, for the errors $\varepsilon(t_\nu)$ of the output values $\eta(t_\nu)$. W.r.t. the computation of such quantities with a reasonable effort, considerable progress has been made in recent years (see e.g. [3] and [4]).

From a practical point of view, however, it seems more important that the user of a package can *specify* the *accuracy level* of the computation. The package should then deliver approximate values consistent with that accuracy specification - or it should give an error message. Virtually all recent ODE-solvers have been built with this design objective.

Generally, the accuracy specification consists in the specification of one *tolerance parameter* δ (the extension of the following considerations to tolerance vectors or tolerance functions presents no principipal difficulties). This quantity δ is the central control parameter for the activities of the package; it thus takes the role which the (constant) stepsize h played in classical discretization methods. It is therefore desirable that δ also be the central parameter in the theoretical analysis of ODE-packages.

In the following, *round-off error* will not be taken into account. It is thus assumed that the arithmetic accuracy of the computation is high enough to keep the effects of round-off errors negligible in comparison with the specified accuracy level. Obviously this could be achieved by having also the wordlength in the computation controlled by the tolerance parameter δ in a suitable manner.

3. T-convergence

What possible meaning can a tolerance parameter δ assume in the performance of a one-pass general purpose ODE-solver? δ can certainly not relate immediately to the *size* of the global error ε: Even if a realistic bound or estimate for ε is generated along with the approximate solution η, there exists no possibility to control the future development of ε by a local strategy.

On the other hand, δ should control $\varepsilon(t)$ in the following sense: Within a reasonable interval of δ-values, the processing of the same ODE under the control first of δ_1 and then of $\delta_2 = r\delta_1$ should lead to global errors $\varepsilon(t;\delta_1)$ and $\varepsilon(t;\delta_2)$ satisfying

$$\varepsilon(t;\delta_2) \approx r\,\varepsilon(t;\delta_1) \tag{1}$$

along the entire path of the integration. If δ is to have any quantitative meaning at all w.r.t. the global error, it must - in my opinion - be the one expressed in (1).

In the following, we will abuse the notation $o(\delta)$ to mean that this term is numerically negligible compared with terms of order δ in the same equation, for values of δ which are reasonable in the particular context; if the $o(\delta)$-term is actually a function of t we will assume that its (piecewise) derivative w.r.t. t is also an $o(\delta)$-term. With this notation, (1) is "equivalent" to

$$\varepsilon(t;\delta) = v(t)\delta + o(\delta), \tag{2}$$

where v is independent of δ and v and v' are bounded on [O,T].

This fundamental property of an ODE-package which properly reacts to the setting of its tolerance parameter δ one could call *tolerance convergence* or *T-convergence*. The essential aspect of (2), however, is not the fact that $\varepsilon \to 0$ as $\delta \to 0$ but the (near-)*proportionality* of ε and δ; hence some suitable translation of the German term "Toleranztreue" might be more suggestive. Note that T-convergence is a property of an ODE-package; it refers to applications of the package to a fixed given problem (0) with varying values of δ.

The validity of (1) or (2) offers to the user the possibility to assess the actual error (componentwise for a system (0)) by performing a second run with a different value of δ (say $r\delta$); he can then assume that the global error $\varepsilon(t;\delta)$ of his original computation satisfies

$$(3) \qquad \varepsilon(t;\delta) \approx \frac{1}{r-1}[\eta(t;r\delta) - \eta(t;\delta)].$$

For $r = 10$, e.g., $\eta(t;\delta)$ would have approximately one significant digit more than $\eta(t;10\delta)$; thus the accuracy of $\eta(t;\delta)$ could easily be judged by a comparison of $\eta(t;\delta)$ and $\eta(t;10\delta)$. In many cases this will give sufficient insight into the qualitative and quantitative behaviour of the global error. (See also a discussion of this fact in [2].)

4. T-consistency

We now want to establish in which manner the tolerance parameter δ has to steer the local action of the ODE-package in order that (1) resp. (2) be valid. In agreement with many publications on the subject we define the *local error* L_n of our computation as follows:

Assume that the computation has proceeded until t_{n-1} (and that $T \gg t_{n-1}$ so that the choice of the next gridpoint t_n is not influenced by the end of the interval). Assume further that t_n and a particular procedure have somehow been chosen and that $\eta(t_n)$ has actually been computed. At the same time, consider $z_n(t)$, the local true solution of the ODE, defined by

$$(4) \qquad z_n'(t) - f(t,z_n(t)) = 0, \quad z_n(t_{n-1}) = \eta(t_{n-1}).$$

We then define the local error L_n by

$$(5) \qquad L_n := \eta(t_n) - z_n(t_n);$$

thus L_n is the error *newly generated* in stepping from t_{n-1} to t_n. In virtually all ODE-packages, an approximate value of L_n is computed and in some way compared with δ in each step.

Furthermore, we observe that the T-convergence requirement (2) really comprises two stages: At first, $\varepsilon(t_n)$ must be in agreement with the size of δ at the gridpoints t_n. Secondly, the interpolation process which defines $\eta(t)$ and $\varepsilon(t)$ for $t \in (t_{n-1}, t_n)$ must also be in agreement with (2).

The grid $\mathbb{G}(\delta)$ which arises, for a given ODE, under the control of δ may for a while remain (nearly) unchanged with decreasing δ (as long as the package can meet higher accuracy requirements by choosing higher order procedures). However, if we include sufficiently small values of δ in our consideration then the grid must eventually become finer as δ continues to decrease. Hence - with our vague usage of the notation o -

(6) $$\text{maximal step in } \mathbb{G}(\delta) = o(1) \quad \text{as } \delta \text{ becomes small.}$$

<u>Theorem 1</u>: If and only if - in the treatment of a fixed problem (O) - the local errors $L_n(\delta)$ generated for a tolerance parameter δ satisfy

(7) $$L_n(\delta) = \overline{\gamma}(t_n, t_{n-1})\ (t_n - t_{n-1})\delta + o(\delta),$$

where $\overline{\gamma}(t,\tau)$ behaves like an integral mean over $[\tau,t]$ of a function which is independent of δ and bounded on $[0,T]$, then (2) is satisfied for all $t_n \in \mathbb{G}(\delta)$.

Proof: a) At first we show that (2) is equivalent to

(8) $$\eta'(t;\delta) - f(t,\eta(t;\delta)) = d(t;\delta), \quad \eta(0;\delta) = y_0,$$

or, equivalently, to

(9) $$\varepsilon'(t;\delta) - f_y(t,y(t))\ \varepsilon(t;\delta) = d(t;\delta) + o(\delta), \quad \varepsilon(0;\delta) = 0,$$

where

(10) $$d(t;\delta) = \gamma(t)\delta + o(\delta)$$

and γ is independent of δ and bounded on $[0,T]$.

That (9) + (10) imply (2) is straightforward, with v being the solution of

(11) $$v' - f_y(t,y(t))v = \gamma(t), \quad v(0) = 0.$$

(8) + (10) imply (9) under our smoothness assumptions.

On the other hand, (2) implies

(12) $$\varepsilon'(t;\delta) - f_y(t,y(t))\varepsilon(t;\delta) = [v'(t) - f_y(t,y(t))v(t)]\delta + o(\delta)$$

which is (9) + (10). Similarly,

$$\begin{aligned}\eta'(t;\delta) - f(t,\eta(t;\delta)) &= [y(t) + \varepsilon(t;\delta)]' - f(t,y(t) + \varepsilon(t;\delta)) \\ &= \varepsilon'(t;\delta) - [f(t,y(t) + \varepsilon(t;\delta)) - f(t,y(t))] \\ &= \quad (12) \quad + o(\varepsilon) = \quad (12) \quad + o(\delta).\end{aligned}$$

b) We now show that the hypothesis of the theorem is equivalent to (8) +

(10):

From (8), (10), and (4) we have for $\varepsilon_n(t;\delta) := \eta(t;\delta) - z_n(t;\delta)$

$$\varepsilon_n'(t;\delta) - [f(t,\eta(t;\delta)) - f(t,z_n(t;\delta))] = \gamma(t)\delta + o(\delta), \quad \text{or}$$

$$\varepsilon_n'(t;\delta) - f_y(t,y(t))\varepsilon_n(t;\delta) = \gamma(t)\delta + o(\delta) + o(\varepsilon_n) + O(\varepsilon)O(\varepsilon_n)$$

and $\varepsilon_n(t_{n-1};\delta) = 0$. From this we obtain (with (2))

$$L_n(\delta) = \varepsilon_n(t_n;\delta) = \delta \int_{t_{n-1}}^{t_n} \Gamma(t_n,\tau)\gamma(\tau)d\tau + o(\delta)$$

$$= \left[\frac{1}{t_n - t_{n-1}} \int_{t_{n-1}}^{t_n} \Gamma(t_n,\tau)\gamma(\tau)d\tau\right](t_n - t_{n-1})\delta + o(\delta).$$

Here $\Gamma(t,\tau)$ is the Green's function of the variational equation associated with (0) and

$$\bar{\gamma}(t_n,t_{n-1}) = \frac{1}{t_n - t_{n-1}} \int_{t_{n-1}}^{t_n} \Gamma(t_n,\tau)\gamma(\tau)d\tau \tag{13}$$

satisfies the conditions of the theorem.

Conversely, since we are not considering intermediate points we may assume from (4) and (5) that η is specified in $[t_{n-1},t_n]$ by

$$\eta(t;\delta) = z_n(t;\delta) + \frac{t - t_{n-1}}{t_n - t_{n-1}} L_n(\delta).$$

Then, in (t_{n-1},t_n),

$$\eta'(t;\delta) = z_n'(t;\delta) + \frac{L_n(\delta)}{t_n - t_{n-1}} = f(t,z_n(t;\delta)) + \frac{L_n(\delta)}{t_n - t_{n-1}}$$

$$= f(t,\eta(t;\delta)) + \bar{\gamma}(t_n,t_{n-1})\delta + O(t - t_{n-1})\delta + o(\delta).$$

With (6) and the conditions on $\bar{\gamma}$, this becomes (8) + (10). □

<u>Theorem 2</u>: If, under the hypothesis of Theorem 1, the values of η between gridpoints satisfy, for some $q_n \geq 1$,

$$\eta(t;\delta) = z_n(t;\delta) + \left(\frac{t - t_{n-1}}{t_n - t_{n-1}}\right)^{q_n} L_n(\delta) + o(\delta), \quad t \in (t_{n-1},t_n), \tag{14}$$

then (2) is satisfied for all $t \in [0,t_N]$ where t_N is the largest t_n in $\mathbb{G}(\delta)$.

<u>Proof</u>: $e_n(t;\delta) := z_n(t;\delta) - y(t)$ satisfies for $t \in [t_{n-1},t_n]$

$$e_n(t;\delta) = w(t)\delta + o(\delta)$$

where

$$w'(t) - f_y(t,y(t))w(t) = 0, \quad w(t_{n-1}) = v(t_{n-1}). \tag{15}$$

A comparison of (15) and (11) shows that (cf. (6))

$$w(t) - v(t) = O(t-t_{n-1}) = o(1) \quad \text{for } t \in (t_{n-1},t_n] \text{ as } \delta \to 0$$

and hence $e_n(t;\delta) = v(t)\delta + o(\delta)$. Thus, by (14) and (7),

$$\begin{aligned} \varepsilon(t;\delta) &= \eta(t;\delta) - y(t) \\ &= e_n(t;\delta) + \left(\frac{t-t_{n-1}}{t_n-t_{n-1}}\right)^{q_n-1} \bar{\gamma}(t_n,t_{n-1})(t-t_{n-1})\delta + o(\delta) \\ &= v(t)\delta + o(\delta) \text{ in } [t_{n-1},t_n]. \end{aligned}$$

We will call an ODE-package *T-consistent* for a given ODE if the hypotheses of both Theorem 1 and 2 are satisfied.

5. Remarks and Observations

Theorem 1 shows that - if T-convergence is accepted as a design objective - then the control mechanism of an ODE-package must attempt to satisfy (7). This means that δ has to control the *error-per-unit-step* and not the error-per-step. Arguments like those contained in section 4 have been put forward previously in a more informal manner in favour of controlling the error-per-unit-step.

Also, eq. (8) + (10) show that in the case of T-convergence the approximate solution η produced by the package is the *exact solution* of an ODE which is a slightly perturbed version of (0), the perturbation being (approximately) proportional to the tolerance parameter δ. Actually, $d(t;\delta)$ in (8) is the *residual* or defect of the numerical solution η in the original problem (0). Thus, δ can be related to a *data error* in the original problem.

Of course, δ would only have an immediate meaning in this context if some suitable measure of $\gamma(t)$ in (10) were equal to unity (or the magnitude of f for relative perturbations). From the point of view of the user of the package this should be a highly desirable feature. In fact, the user might even like to be able to specify the weighting of the components of his system for the purpose of error control. Such a facility has been made available in the ODE-package described in [2].

The following is a clever way of implementing a control mechanism of type (7): Often, in the course of the computation of $\eta(t_n)$ an intermediate value $\tilde{\eta}(t_n)$ appears whose accuracy is "one order lower" in terms of classical discretization theory. This is regularly the case in predictor-corrector type methods, furthermore in one-step methods of the England/Fehlberg type where the lower and higher order values are produced systematically for the purpose of local error estimation. In this case,

$$(16)\qquad \begin{cases} L_n(\delta) = \eta(t_n;\delta) - z_n(t_n;\delta) \approx (t_n - t_{n-1})^{p+1}\varphi_n, \\ \tilde{\eta}(t_n;\delta) - z_n(t_n;\delta) \approx (t_n - t_{n-1})^{p}\,\tilde{\varphi}_n, \end{cases}$$

where φ_n and $\tilde{\varphi}_n$ are values of the principal error function of the procedure involved.

If we may safely assume that the η-value is considerably more accurate than the $\tilde{\eta}$-value (in terms of local error) so that (cf. also (6))

$$(17)\qquad \tilde{\eta}(t_n;\delta) - \eta(t_n;\delta) = \tilde{\eta}(t_n;\delta) - z_n(t_n;\delta) + o(\delta)$$

then the control requirement

$$(18)\qquad \tilde{\eta}(t_n;\delta) - \eta(t_n;\delta) \approx \tilde{\gamma}(t_n,t_{n-1})\delta,$$

where $\tilde{\gamma}$ satisfies the same conditions as $\overline{\gamma}$ in (7), implies, by (16) and (17),

$$L_n(\delta) \approx (t_n - t_{n-1})\frac{\varphi_n}{\tilde{\varphi}_n}\tilde{\gamma}(t_n,t_{n-1})\delta,$$

which is of type (7). Thus, although the control requirement (18) looks like an error-per-step criterion it actually has the effect of an error-per-unit-step control upon η since it is not the local error of η but that of $\tilde{\eta}$ which is controlled by (18).

This trick seems to have been deliberately used for the first time in [2] where also the above reasoning is given to justify its use. Note that the numerical solution must be continued with the (more accurate) η-value which is what one would like to do anyway. Of course, φ_n and $\tilde{\varphi}_n$ are normally inaccessible so that no normalization of $\overline{\gamma}$ can be effected when one uses this trick.

Two serious arguments may be brought forward against the error-per-unit-step control (7):

1) Jump discontinuities in f: If f happens to have a jump discontinuity at $\overline{t}$ which is not explicitly dealt with by the driver program then normally $L_n \approx (t_n - \overline{t})\overline{c}$ in a step straddling $\overline{t}$ where $\overline{c}$ is rather independent of the procedure used. Hence, if t_{n-1} is close to $\overline{t}$ - which is very likely to be the case due to the error control effort in the previous step - then (7) cannot be satisfied by a reduction of $t_n - t_{n-1}$ (except when $\overline{c} < \overline{\gamma}\delta$) and the computation gets stuck.

This situation may be overcome by using a control of type (18) which naturally will not be a consistent criterion in the above situation but normally carry the computation across the discontinuity with a reasonable stepsize.

2) Stiff systems: As Lindberg has shown in [5], an error-per-unit-step control with a constant $\bar{\gamma}$ will generally produce a step sequence much inferior to the one generated by a naive error-per-step control while the computation runs through the transient phase of the computation. However, his analysis also reveals the reason and shows how one should theoretically choose (and practically estimate) a $\bar{\gamma}$-function in (7) which will make the error-per-unit-step control efficient.

Furthermore, Lindberg has pointed out in [5] that in the stationary phase of the computation the error e_n of the local solution z_n may be damped so strongly that the local error $\eta(t_n) - z_n(t_n)$ becomes virtually equal to the global error $\eta(t_n) - y(t_n)$. It can be seen rather easily from our theory how this should effect (7): According to the proof of Theorem 1, $\bar{\gamma}$ should be of the form (13). If $\| \Gamma(t_n,\tau) \| = \text{const}\; e^{-M(t_n-\tau)}$, with $M > 0$ very large, we obtain $\bar{\gamma}(t_n,t_{n-1}) = O\left(\frac{1}{t_n - t_{n-1}}\right)$ and (7) becomes an error-per-step control.

6. Further requirements

From section 4 it would seem that T-consistency is not only necessary but also sufficient for T-convergence, at least for sufficiently smooth ODE's. The non-appearance of an explicit stability requirement for the procedures used in the package may be explained by the fact that the control mechanism (7) enforces the generation of a solution of the perturbed ODE (8). Thus no stability problem can arise for a well-behaved ODE and a small perturbation.

However, there must be a snag; it is revealed by the consideration of a two-step Hermitian extrapolation method under the control of (7): In order to satisfy (7), the step selection mechanism will have to decrease the steplengths in a *geometric progression* and the computation will "die" after a short interval. This observation does not invalidate Theorems 1 and 2 whose assertions refer only to an interval $[0,t_N]$; but we need the further requirement that $t_N \geq T$ for all sufficiently well-behaved ODE's and δ from a reasonable range.

A formalization of this requirement has to deal with various non-trivial aspects of this situation: E.g., a strong local decrease of the stepsizes may be perfectly legitimate when it stems from a sudden change in the behavior of the ODE; also the steplengths will only be permitted to change in a restricted fashion in many packages.

A possible way of dealing with the influence of the problem may be the following one which is suggested by the well-known handling of the same problem in the theory of D-stability: One might be able to prove that no geometric deterioration of the step sequence can occur for a sufficiently smooth ODE if it cannot occur in the application of the same package to $y' = 0$. A theorem of this type would reduce the analysis to the consideration of the situation for this simple ODE.

But the situation is still non-trivial: Somehow we will have to ascertain that the package cannot run into geometric deterioration when started from an arbitrary "admissible" sequence of t_n and $\eta(t_n)$ values, i.e. a sequence which *could* have been generated by the package working on $y' = 0$ under tolerance δ. The quantitative characterization of these admissible sequences is not immediately clear; if we can prove that the choice of $\bar{\gamma}$ in (7) is irrelevant we could simply require

$$|\nabla\eta_n| := |\eta(t_n;\delta) - \eta(t_{n-1};\delta)| \leq (t_n - t_{n-1})\delta .$$

However, the dependence of $\nabla t_n := t_n - t_{n-1}$ and of $\nabla\eta_n$ on $\{\nabla t_{n-\kappa}, \nabla\eta_{n-\kappa}\}_{\kappa=1(1)k}$ will be non-linear and discontinous for most ODE-solvers even with $y' = 0$; thus we can - if at all - use a "root criterion" only after further structural simplification. Also one may have to take into account that the step selection mechanism does not know the correct local error but only an estimate, which may be crucial in the case of instability.

Example: Two-step Hermitian extrapolation, $y' = 0$:

Take $\eta(t_{n-1};\delta) = 0$, $\eta(t_{n-2};\delta) = (t_{n-1} - t_{n-2})\delta$, then the requirement

$$L_n = \eta(t_n;\delta) - 0 = (t_n - t_{n-1})\delta$$

leads to $t_n - t_{n-1} \approx .28(t_{n-1} - t_{n-2})$ independently of δ and thus to fast geometric deterioration.

If we do not use the true value of L_n but its estimate obtained by a comparison of $\eta(t_n;\delta)$ with a corrector value and if we agree to reduce h only by powers of 2 we obtain the following story:

1st attempt: $t_n - t_{n-1} = t_{n-1} - t_{n-2}$, est.$L_n = 4\,(t_n - t_{n-1})\delta \Rightarrow$ reject

2nd attempt: $t_n - t_{n-1} = \frac{1}{2}(t_{n-1} - t_{n-2})$, est.$L_n = \frac{27}{16}(t_n - t_{n-1})\delta \Rightarrow$ reject

3rd attempt: $t_n - t_{n-1} = \frac{1}{4}(t_{n-1} - t_{n-2})$, est.$L_n = \frac{25}{32}(t_n - t_{n-1})\delta \Rightarrow$ accept.

Both situations are rather simple; but this changes as soon as we regard methods which use 3 or more backward points in the computation of $\eta(t_n)$.

For ODE-solvers whose procedures reduce to *one-step* procedures for $y' = 0$, the above-mentioned theorem would be all that is needed. This would include Adams-PC-type packages and RK-type packages which account for most of the presently used ODE-solvers; it would not suffice for

GBS-packages or for Gear-type packages using backward differentation schemes.

Theoretically, it would also be sufficient if a package would "retreat" to one-step procedures if it experienced bad difficulties; most packages do just that.

7. Experiments

Some preliminary numerical experiments were run to verify the T-convergence property (2) for two well-known ODE-packages:

a) the Shampine-Gordon package (Adams PC) as described in [2];
b) the GBS-code DIFSY1, with a step and extrapolation control as described in [6].

Both packages were applied to a set of 36 test ODE's taken from well-known sources; the set did not contain badly stiff problems. Each test problem was solved over its associated interval [O,T] with 9 different values of the tolerance parameter:

$$\delta_m = 10^{-m}, \quad m = 2(1)10.$$

(The computer was a CDC CYBER 74 where round-off may still be considered negligible at a tolerance level 10^{-10}.)

For each test problem, the $\varepsilon(T;\delta_m)$, $m = 2(1)10$, were found from the known exact solution and a straight line

$$\log \varepsilon(T;\delta) = s \log \delta + \log c \tag{19}$$

was fitted by least squares through the point set $\{\log \delta_m, \log \varepsilon(T;\delta_m)\}$. Since (19) corresponds to

$$\varepsilon(T;\delta) = c\, \delta^s,$$

s should come out close to 1 and the standard deviation of the fit should be small in the case of T-convergence.

Here are the results of a first evaluation of the test runs which did not include the computation of the standard deviation:

fitted s was between		for		
.98 and	1.02	7	2	
.95	1.05	15	9	
.90	1.10	26	16	of the 36 test ODE's
.85	1.15	31	21	
.825	1.175	33		
.80	1.20	33	24	
.70	1.30	33	29	
	with package	a)	b)	

These preliminary figures show that the Shampine/Gordon package may be called T-convergent for a wide range of ODE's from a practical point of view. Of the 3 remaining test problems which led to s-values outside [·825, 1·175], 2 had low-order (piecewise) polynomial solutions and were solved "exactly" for all values of δ while the third had a random generated term in it and would thus not be covered by any simple-minded theory. Also the various s-values were nicely distributed around 1, with 22 larger and 14 smaller than 1. This "proves" that the trick described in section 5 (cf. eq. (16) to (18)) actually led to an error-per-unit-step control of type (7); for an error-per-step control the s-values should have been consistently smaller than 1.

The main reason for the considerably "poorer" performance of the GBS-package was that it rather consistently produced too high an accuracy for low δ-values, which leads to fitted s-values below 1. Actually, for this package only 7 test problems yielded an $s > 1$ while the other 29 produced an $s < 1$, often considerably smaller. On the other hand, the values of c were much smaller than with package a) for most of the test ODE's. It will be interesting to see from further experiments whether (2) is more closely satisfied when the range of δ-values is shifted towards smaller values.

It is planned to extend the experiments in various ways:

- evaluation of the standard deviation in the fitting of (19),
- variation of the -ranges,
- variation of the output abscissas,
- inclusion of other packages.

The results of these experiments will be reported in detail after their completion.

8. Conclusion

In this report, I have attempted to outline an analysis of ODE-packages with the tolerance parameter δ as its central and only parameter. A good deal of this analysis has been rather superficial and incomplete; it will need further elaboration and refinement. Perhaps the concepts of T-convergence and T-consistency should even be replaced by different ones; certainly the names for these concepts are rather unfortunate and more suggestive names should be found.

On the other hand, I strongly feel that basically this is the right approach to be used in the analysis of general purpose one-pass ODE-solvers, and I would hope that this report will stimulate further research in this direction.

Literature

[1] H.J. Stetter, Analysis of discretization methods for ordinary differential equations, Springer, 1973.

[2] L.F. Shampine - M.K. Gordon, Computer solution of ordinary differential equations: the initial value problem, Freeman, 1975.

[3] U. Marcowitz, Fehlerabschätzung bei Anfangswertaufgaben für Systeme gewöhnlicher Differentialgleichungen mit Anwendung auf das Reentry-Problem, Numer. Math. 24 (1975) 249 - 275.

[4] L.F. Shampine - H.A. Watts, Global error estimation for ordinary differential equations, ACM - TOMS 2 (1976) 172 - 186.

[5] B. Lindberg: Optimal stepsize sequences and requirements for the local error for methods for (stiff) differential equations, TR 67, Comput. Science, Univ. of Toronto, 1974.

[6] H.G. Hussels, Schrittweitensteuerung bei der Integration gewöhnlicher Differentialgleichungen mit Extrapolationsverfahren, Diplomarbeit, Universität Köln, 1973.

BOUNDARY VALUE PROBLEMS IN INFINITE INTERVALS

J. Waldvogel
Seminar fuer Angewandte Mathematik
ETH-Zentrum, CH-8092 Zurich

Abstract

The treatment of differential equations on infinite intervals (assuming that solutions exist on the entire interval)depends crucially on the asymptotic behaviour of the solutions at infinity. Often, power series with logarithmic coefficients solve the differential equations formally and may be proven to be asymptotic to true solutions.

In this paper it is suggested to use numerical integration over a finite portion of the interval, and to switch to the asymptotic approximation when the latter is sufficiently accurate. In order to integrate numerically over large intervals a step roughly proportional to the value of the independent variable may often be used. One-step methods with step control are well suited for this purpose.

The shooting method is applied in order to solve boundary value problems. If numerical integration is done in the stable direction, the secant method is an efficient tool for determining missing initial conditions. Bisection is preferrably used when integrating in an instable direction.

These methods are illustrated with a practical example.

1. INTRODUCTION

An interval that extends to infinity on one or both sides reflects the rather frequent physical situation of virtually unlimited space. Whereas truncating the infinite interval may be an adequate approximation in many practical cases, the mathematical problem of handling infinite intervals is interesting in itself. Furthermore,differential equations on infinite intervals are to be solved frequently in boundary layer methods.

Here an algorithm combining analytical and numerical methods for solving asymptotic boundary value problems will be proposed. It consists of a modified shooting method where the trial solutions are initialized by asymptotic expansions at infinity and continued by numerical integration. Often, a step size roughly proportional to the independent variable may be used. Then, the secant method efficiently selects the solution satisfying the boundary conditions.

Compared to other methods (e.g. homotopy between the infinite interval and a finite portion) the present approach has the advantage of maximum usage of analytic information, which is often not too difficult to obtain. As a consequence, high accuracy may be obtained if a good numerical integrator is available. On the other hand, the method is basically restricted to analytic differential equations, and a completely automatic application seems to be difficult.

2. FORMAL SOLUTIONS

The theory of formal (possibly divergent) series solutions and asymptotic solutions of nonlinear analytic differential equations is of considerable complexity (Wasow, 1965; Coddington and Levinson, 1955). The existence of asymptotic solutions of certain forms depends crucially on the detailled structure of the right-hand sides. Going back to the corresponding theorems exceeds the scope of this work. We shall restrict ourselves to describing two rather heuristic methods that will readily produce formal solutions in most cases. Sometimes it follows directly from the relevant theorems that such formal series are asymptotic to true solutions as $x \to \infty$.

We consider the system

$$(1) \qquad \dot{y} = f(x, y), \qquad \dot{} \equiv \frac{d}{dx}$$

of first order differential equations, where $y(x)$ and $f(x,y)$ are n-vector and f is analytic in each component of y. According to the most general results Equ.(1) has formal series solutions

$$(2) \qquad y(x) = Y(x^{\alpha}, (\log \frac{x}{x_0})^{\beta}, e^{\gamma x^{\delta}}),$$

where the arguments of the multiple Taylor series Y may occur repeatedly with different values of α, β, γ, δ. The formal solution of an n-th order system is said to be <u>complete</u> if it contains n independent parameters.

A first method for solving differential equations by series valid for $x \to \infty$ is <u>formal Picard iteration.</u> Beginning with an appropriate initial guess $y_0(x)$ for the solution $y(x)$, we iterate

$$(3) \qquad y_{j+1}(x) = \int f(x, y_j(x))\,dx, \quad j = 0, 1 \ldots$$

where the indefinite integration is carried out by expansion with respect to decreasing powers of x. Hereby n integration constants are introduced. If these parameters can be chosen such that each iteration step produces additional correct terms in the series, the iteration is said to be formally convergent. This process yields the complete formal solution.

Throughout the paper we shall use one particular example in order to illustrate the principal ideas.

Example:

$$\ddot{y} = \frac{-x}{(x^2+1)(y^2+1)} \quad . \tag{4}$$

Let $y_0(x) = O(x)$ be the initial "approximation" in the formal Picard iteration. By repeatedly inserting this into the right-hand side of (4) and integrating twice we obtain

$$\begin{aligned} y_1 &= ax + b + O(x^{-1}) \\ y_2 &= ax + b - \frac{1}{2a^2} x^{-1} + \frac{b}{3a^3} x^{-2} + O(x^{-3}) \quad \text{etc.,} \end{aligned} \tag{5}$$

where a and b are integration constants.

Each iteration yields two additional terms of the formal series. It follows, however, that we have to assume $a \neq 0$; hence a one-dimensional subfamily of formal solutions is missing from the two-dimensional family (5), namely the family of solutions with $\dot{y}(\infty) = 0$.

In order to find the formal series in this case we propose a "trial and error" algorithm as an alternative method. We attempt to directly establish a series of the form (2) formally satisfying the differential equation (1). The simplest possibility, a Laurent series with unknown coefficients, is tried first. Inserting the series into (1) results in a system of equations for these coefficients (a nonlinear equation for the leading coefficient and linear equations for the higher coefficients). If a solution of these equations with at most n parameters can be found, the formal series solution exists.

Failure of this algorithm is often indicated by the vanishing of the determinant in a system of linear equations. Then the formal solution can still be established in the same form if the coefficients of the series are assumed to be polynomials in

$$\ell = \log \frac{x}{x_0} = \log x + \ell_o \ , \tag{6}$$

where x_o or ℓ_o are free parameters.

If this device fails one tries, according to (2), to introduce fractional exponents, exponential terms, etc.

Example:

The solutions of Equ. (4) with $\dot{y}(\infty) = 0$ may be found by assuming

(7) $$y(x) = cx^{\alpha} + o(x^{\alpha}), \quad \alpha > 0$$

with unknown real quantities c, α. Inserting this into (4) yields

$$c\,\alpha(\alpha-1)\,x^{\alpha-2} + o(x^{\alpha-2}) = -c^{-2}x^{-1-2\alpha} + o(x^{-1-2\alpha}),$$

whence

(8) $$\alpha = \frac{1}{3}, \quad c = \sqrt[3]{4.5} \; .$$

Next, it is reasonable to assume the formal solution to be a Laurent series in $x^{1/3}$, i.e.

(9) $$y = cx^{1/3} + b_o + b_1 c^{-1} x^{-1/3} + \ldots .$$

By substitution into (4) the conditions

$$-\frac{2}{9}c^3 = -1$$

$$-\frac{4}{9}c^2 b_o = 0$$

$$\frac{4}{9}c\,b_1 - \frac{4}{9}c\,b_1 - \frac{2}{9}(b_o^2 + 1) = 0$$

are obtained, which uniquely result in $c = (9/2)^{1/3}$, $b_o = 0$ in agreement with (8). There is no value of b_1, however, satisfying the third condition; hence a formal solution of the form (9) does not exist. The series (9) can still be established, however, if its coefficients are allowed to be polynomials in ℓ. If we omit the even terms from the series (since b_o, b_2, b_4, ... will turn out to vanish) we have

(10) $$y(x) = cx^{1/3} + b_1(\ell)\,c^{-1}x^{-1/3} + b_3(\ell)c^{-3}x^{-1} + o(x^{\varepsilon-5/3})$$

instead of (9). By using the differentiation rule

$$\frac{d^2}{dx^2}\{x^{\alpha}\,b(\ell)\} = x^{\alpha-2}\{b'' + (2\alpha-1)b' + \alpha(\alpha-1)\,b\} \qquad ' \equiv \frac{d}{d\ell}$$

the conditions for determing the polynomials $b_j(\ell)$ become themselves differential equations:

$$(11)\qquad b_1'' - \frac{5}{3} b_1' = \frac{2}{9}$$

$$(12)\qquad b_3'' - 3b_3' + \frac{14}{9} b_3 = -\frac{2}{9}(3b_1^2 + 4b_1 + 1) \qquad \text{etc;}$$

in general

$$(13)\qquad b''_{2k+1} - \frac{4k+5}{3} b'_{2k+1} + \frac{k(4k+10)}{9} b_{2k+1} = f_{2k+1} ,$$

where f_{2k+1} is a polynomial in $b_1, b_3, \ldots, b_{2k-1}$.
The differential equation (11) is satisfied by the linear polynomial

$$(14)\qquad b_1(\ell) = -\frac{2}{15}\ell$$

which is not unique; an additive constant ℓ_o has already been included in the definition (6) of ℓ. The other solutions of (11) contain exponential functions and need not be considered. It is essential that the coefficient of b_1 in (11) vanishes; this allows $b_1(\ell)$ to be a <u>linear</u> polynomial, although the right-hand side is a constant.

It remains to be shown that the conditions (13) admit polynomial solutions also for k = 1, 2, ... As an induction hypothesis we assume that there exist polynomials $b_1(\ell), \ldots, b_{2j-1}(\ell)$ satisfying Equ. (13) for k = 0, 1, ...j-1. This is certainly true for k = 0. Then, the right-hand side f_{2j+1} is also a polynomial in ℓ. By the method of unknown coefficients it is seen that Equ. (13) has a unique polynomial solution for $k = j > 0$. This is a consequence of the nonzero coefficients of b_{2k+1}. Herewith the induction is complete.

The second step of this recursive process yields from Equ. (12):

$$(15)\qquad b_3(\ell) = -\frac{4}{525}\ell^2 + \frac{172}{3675}\ell - \frac{367}{8575} .$$

3. NUMERICAL INTEGRATION

For representing solutions of differential equations in infinite intervals we suggest to use an asymptotic formula (mostly a truncated formal series) as an approximation to the solution for $x \geq X$, where the (large) number X is chosen such that the asymptotic formula at x = X yields sufficient accuracy. Beginning with initial conditions (obtained from the asymptotic formula) at x = X, the solution may then be continued by backwards numerical integration.

Basically, numerical integration should be done in the direction of slowest error growth. This may be determined from the variational equation. If forward integration is preferrable, the numerical solution may be matched to the asymptotic solution at $x = X$. A severe loss of accuracy may occur if the error grows exponentially in both directions. Numerical integration over large intervals can be a formidable task. However, if the solutions are sufficiently smooth as $x \to \infty$, a step size h roughly proportional to x, i.e. $h = \varepsilon x$ with a constant ε may often be used. Therefore, as $x \to \infty$ the number of steps in every "decade" $a \leqslant x < 10\,a$ is roughly the same.

One-step methods with automatic step size control are well suited for numerical integrations of this type. Our experiments were carried out with Fehlberg's (1969) combined Runge-Kutta methods of orders 7 and 8 (13 function evaluations). The step control mechanism automatically chose an approximate geometric sequence for the steps.

Example:

We shall briefly discuss the calculation of the particular solution of (4) with the asymptotic behaviour (10), given by a specific value of the parameter ℓ_0. The variational equation

$$(16)\qquad \ddot{\eta} = \frac{2xy}{(x^2+1)(y^2+1)^2}\,\eta$$

of (4), taken at a reference solution $y(x) \sim cx^{1/3}$ has the solutions

$$(17)\qquad \eta_1 = O(x^{4/3}), \qquad\qquad \eta_2 = O(x^{-1/3}).$$

Hence the error grows slowest in backwards numerical integration. The slight instability even in this direction results in a modest, still acceptable, loss of accuracy.

The relative error in the asymptotic formula (10) is of order $O(x^{-2})$. Therefore, with $X = 10^6$ we can expect an accuracy of about 12 significant figures in the initial values at $x = X$.

For various tolerances limiting the local relative truncation error the following numbers of steps per decade were carried out by the Runge-Kutta-Fehlberg integrator:

tolerance	10^{-8}	10^{-9}	10^{-10}	10^{-11}	10^{-12}	10^{-13}	10^{-14}
steps per decade	6	8	11	15	21	28	38.

With the tolerance 10^{-13} backwards integration from $x = 10^6$ to $x = 0$

needed a total of 187 steps.

4. SHOOTING METHODS

Shooting methods solve boundary value problems by iteratively determing the missing initial conditions at one of the boundaries. The most serious difficulty with this method is the loss of accuracy far away from this boundary. This problem is even more serious when dealing with large or infinite intervals. A common remedy is to resort to multiple shooting techniques (see, e.g., Stoer, Bulirsch, 1973).

Here we shall consider only the simple shooting method. In cases of stability or modest instability the unknown initial conditions are determined by means of the secant or the modified Newton iteration. If the error grows exponentially, bisection is probably better suited.

Example:

The boundary value problem $y(0) = 0$, $\dot{y}(\infty) = 0$ of Equ. (4) will be considered. According to section 3 shooting will be done from $x = X = 10^6$ back to $x = 0$. Here the quantity corresponding to the missing initial condition is the parameter ℓ_o which enters via Equ. (6) into the asymptotic formula (10). The condition to be met at $x = 0$ is $y(0) = 0$.

By initializing the secant method with the two values $\ell_o^{(0)} = 0$, $\ell_o^{(1)} = 10$ and by using the tolerance 10^{-13} for numerical integration the following results are obtained:

i	ℓ_o	y(0)	$\dot{y}(0)$
1	10	.11683 85901	.86792 75841
2	11.04226 153	-.04360 84622	.93009 71640
3	10.75898 165	.00107 96512	.91281 22977
4	10.76582 560	.00000 95470	.91322 69085
5	10.76588 666	-.00000 00018	.91323 06080
6	10.76588 665	-.00000 00000	.91323 06073

Only the leading 8 figures of ℓ_o influence the initial conditions at $x = X = 10^6$ in these 14-digit calculations. Therefore the last two figures of ℓ_o are not reliable. However, based on calculations with different tolerances and different values of X, the last value of $\dot{y}(0)$ is believed to be correct to 10 decimals. The loss of accuracy due to the slight instability amounts to about 3 figures.

References.

Coddington, E.A. and N. Levinson, 1955: Theory of Ordinary Differential Equations.
Mc Graw-Hill, New York.

Fehlberg, E., 1968: Classical Fifth-, Sixth-, Seventh- and Eighth-Order Runge-Kutta Formulas with Stepsize Control.
NASA TR R - 287.

Stoer, J. und R. Bulirsch, 1973: Einführung in die Numerische Mathematik II.
Springer, Berlin.

Wasow, W., 1965: Asymptotic Expansions for Ordinary Differential Equations.
John Wiley, New York.

Capacitance Matrix Methods for Helmholtz's Equation on General Bounded Regions

Olof Widlund
Courant Institute
251 Mercer Street
New York, N.Y. 10012/USA

1. Introduction. In this paper, we shall give a brief survey of some recent work on capacitance matrix methods for the numerical solution of Helmholtz's equation,

$$-\Delta u + cu = f \quad \text{on } \Omega .$$

On the boundary $\partial\Omega$ of the bounded region Ω either Dirichlet or Neumann data is given. We shall discuss both two and three dimensional problems and, for convenience, restrict ourselves to values of the constant c which make the problem uniquely solvable.

Capacitance matrix algorithms are methods for the solution of the highly structured linear systems of algebraic equations which arise when the Helmholtz problem is discretized by a finite difference method. We can, for example, use the five, in three dimensions seven, point formula and combine it with the Shortley-Weller approximation at those so-called irregular mesh points in the open set Ω which fail to have all their relevant neighbors in Ω, see Collatz [9]. The region Ω is imbedded in a rectangle or another region for which the Helmholtz equation can be solved by separation of the variables and a mesh is introduced on the enlarged simple region. The discrete Helmholtz problem on this simple region should be solvable by a very fast method. This requirement introduces certain intrinsic limitations on the choice of the mesh, see further a discussion in Widlund [35], and these limitations are inherited by the mesh used for the problem on the region Ω. The development of fast, reliable solvers on simple regions has been and remains a very active area of research, see for example Bank [2], Buneman [4], Buzbee, Golub and Nielson [8], Fischer, Golub, Hald, Leiva and Widlund [13], Hockney [19,21], Schröder and Trottenberg [32] and Swarztrauber and Sweet [34]. A main advantage of a properly designed capacitance matrix program is that it can be speeded up, at the expense of a modest programming effort, by the replacement of a subroutine,

*The work presented in this paper was supported by the ERDA Mathematics and Computing Laboratory, Courant Institute of Mathematical Sciences, New York University, under Contract No. E(11-1)-3077 with the Energy Research and Development Administration.

whenever a faster Helmholtz solver becomes available.

Early work on capacitance matrix methods was carried out by Buzbee, Dorr, George and Golub [7]. We follow their presentation of the imbedding idea in Section 2. At about the same time Hockney [20,22] designed a similar method based on an idea from potential theory suggested to him by Oscar Buneman. George [15] also contributed a very interesting idea by showing how methods which do not require the computation of the capacitance matrix could be developed. The method was further developed by Buzbee and Dorr [5,6]. For a detailed discussion of all this work and a much more complete description of two of our own methods see Proskurowski and Widlund [31]. Here we will also report on more recent unpublished work by Banegas [1], O'Leary and Widlund [26], Proskurowski [30] and Shieh [33]. While most of this work deals specifically with second order accurate finite difference schemes, we have also shown, that higher order accuracy can be obtained, see Pereyra, Proskurowski and Widlund [27]. In that study, we combine a capacitance matrix method, a family of difference schemes suggested by Heinz-Otto Kreiss and a deferred correction method to obtain highly accurate solutions. Proskurowski [29] has also combined a capacitance matrix method with a block Lanczos algorithm to design a method for the computation of the eigenvalues and eigenfunctions of the Laplacian on an arbitrary bounded region in the plane.

We note in conclusion that the applicability of our Helmholtz solvers can be extended to other second order elliptic problems by combining them with different generalized conjugate gradient methods, see Bartels and Daniel [3], Concus and Golub [10], Concus, Golub and O'Leary [11], O'Leary [26] and Widlund [36].

2. <u>The imbedding of the discrete Poisson problem</u>. To be specific let us assume that we use Cartesian coordinates, that the problem is two dimensional and that we imbed the region Ω in a rectangle. We further assume that the mesh is uniform with the same mesh width, h, and with m and n mesh points respectively in the two coordinate directions. The boundary conditions on the rectangle can be of arbitrary type but should allow us to use a fast solver. We will however see, in Section 4, that by making the solution periodic we can decrease the cost of computing the capacitance matrix considerably. We will need a frame of exterior mesh points, one mesh width wide, next to the boundary of the rectangle but otherwise the position of the region Ω in the rectangle is of no noticeable importance. Some, but not all, fast solvers work only, or are much more efficient, for certain values of m and n. The choice of

these parameters thus depends on obvious considerations of execution time and storage required by the fast solver.

There are three disjoint sets of mesh points. They are Ω_h, the set of interior mesh points which have all their relevant neighbors in Ω, $\partial\Omega_h$, the set of irregular mesh points introduced in Section 1 and $(C\Omega)_h$, the set of exterior mesh points which belong to the complement of Ω. For all the interior mesh points, we use the basic discretization which matches the fast solver. For the irregular points, we combine a discrete Helmholtz operator with some interpolation formula to approximate the boundary condition. Finally, we extend the linear system of equations by using the same basic formula for the points of $(C\Omega)_h$. The data is extended in an arbitrary way to these exterior points. Most of our programs will also produce largely arbitrary values of a mesh function at these points, a useless by-product of the fast Helmholtz solver.

Let us denote by A the matrix corresponding to our difference equations on the entire mesh. We order the equations and unknowns in the same order as those of the regularly structured problem for which a fast solver is available. We denote this second matrix by B. It is easy to see that the only rows of A and B which differ are those corresponding to the irregular mesh points. We can therefore write

$$A = B + UZ^T,$$

where U and Z have p columns and p is the number of elements of $\partial\Omega_h$. We can choose U to be an extension operator which maps any mesh function defined only on $\partial\Omega_h$ onto a function defined for all mesh points. It is constructed so that the values on $\partial\Omega_h$ are retained while the remaining values are set to zero. Its transpose, U^T, is a trace operator which maps any function defined for every mesh point onto its restriction to $\partial\Omega_h$. The matrix Z^T can be regarded as a compact representation of A-B. We note that the matrices U and Z are quite sparse.

In our formulas for the irregular mesh points values of the mesh function at exterior mesh points should not appear. If we choose a suitable permutation matrix P

$$P^TAP = \begin{pmatrix} A_{11} & 0 \\ A_{21} & A_{22} \end{pmatrix},$$

and we see that A is a reducible matrix. Here the block A_{11} corresponds to the difference equations on the set $\Omega_h \cup \partial\Omega_h$. It is easy to see from this structure that the restriction of the solution u of the equation

$$Au = f , \tag{2.1}$$

to this set of mesh points is independent of the solution and data at the exterior points.

If we assume that A and B are invertible, we can use the Woodbury formula, see Householder [23], and write the solution of equation (2.1) as

$$u = B^{-1}f - B^{-1}U(I + Z^T B^{-1} U)^{-1} Z^T B^{-1} f \ .$$

The $p \times p$ matrix $C = I + Z^T B^{-1} U$ is the capacitance matrix of this variant of the method. The solution of the problem can thus be obtained at the expense of generating the matrix C, factoring C into triangular factors, solving a linear system of equations using these factors and employing the fast Helmholtz solver twice. We note that the first two steps of this process are independent of the data f and need not be repeated if additional problems are to be solved for the same difference equations. We also note that the requirement that A and B be invertible can be relaxed, see Buzbee, Dorr, George and Golub [7] and Proskurowski and Widlund [31].

3. Potential theory and discrete dipoles. Our interest in capacitance matrix methods grew out of an observation of a formal analogy between the Woodbury formula and a classical solution formula for the Neumann problem for Laplace's equation as presented in Courant-Hilbert [12], Garabedian [14] and Petrowsky [28]. A detailed discussion of the material presented in this section is given in Proskurowski and Widlund [31] and in a fine Ph.D. thesis by Shieh [33]. In potential theory the solution of the Neumann problem is written as the sum of a space potential term, which has the effect of reducing the differential equation to one which is homogeneous, and a potential from a single layer charge distribution on the boundary. These terms essentially correspond to those of the Woodbury formula. The charge distribution is determined by solving a Fredholm integral equation of the second kind. The integral operator has the form identity plus compact and it has, for $c = 0$, an inverse of the same form on a space of codimension one. For positive values of the constant c the integral operator is nonsingular. The same Ansatz gives rise to an ill posed problem, a Fredholm integral equation of the first kind, when applied to a Dirichlet problem. This lead us to the conjecture that the capacitance matrices derived in the previous section should be uniformly well conditioned for the Neumann problem if $c > 0$, but increasingly ill conditioned for the Dirichlet case when the value of p increases. This conjecture has been borne out in practice, see Proskurowski and Widlund [31]. We are interested in the condition

number of these matrices primarily because a suitable spectral distribution will allow us to use rapidly convergent iterative methods to solve the capacitance matrix equation.

In a successful Ansatz for the continuous Dirichlet problem the single layer potential is replaced by the potential from a dipole layer. The dipole density is determined by solving a Fredholm integral equation of the second kind. While the Woodbury formula corresponds to a single layer Ansatz of the form

$$u = B^{-1}f + B^{-1}U\rho \ ,$$

we now write the solution of the discrete Dirichlet problem as

$$u = B^{-1}f + B^{-1}V\mu \ .$$

Here μ is a vector with p components. Each column of the matrix V represents a discrete dipole of unit strength. If we regard such a column as a mesh function, it vanishes everywhere except at one irregular mesh point and at two of its neighboring mesh points which lie at a distance h and $\sqrt{2}$ h respectively. These two points are chosen so that an outside normal, originating at the irregular point, lies in the cone spanned by the vectors from the irregular point to the two neighbors. A similar construction is carried out, using four points, in three dimensions. The sum of the charges of any discrete dipole is zero and their relative weights are chosen so as to simulate a dipole in the direction of the normal. With this Ansatz the capacitance matrix equation becomes

$$C\mu = (I + Z^TB^{-1}V)\mu = -Z^TB^{-1}f \ . \tag{3.1}$$

This is a formally convergent approximation to the Fredholm integral equation of the second kind provided certain scale factors are properly chosen.

The solution formula based on the dipole Ansatz is correct if all mesh functions of the form $V\mu$ vanish on Ω_h. In our program for the three dimensional case, we check that this condition is satisfied. It introduces a certain, rather mild, restriction on the type of regions and the mesh sizes which can be handled by our method.

The capacitance matrices resulting from the dipole Ansatz are, in our experience, quite well conditioned with a spectral distribution similar to that of the Fredholm integral operator. For the five point formula, some particular choices of the boundary approximation and a correct way of scaling Shieh [33] has shown that the singular values of the capacitance matrix corresponding to the single layer Ansatz for the

Dirichlet problem converge to those of the Fredholm operator of the first kind. These capacitance matrices must therefore be increasingly ill conditioned. For the single layer Ansatz for the Neumann problem and the discrete dipole Ansatz for the Dirichlet problem he writes

$$C = B_h + K_h .$$

The matrix B_h represents the coupling between the irregular mesh points which are within $\sqrt{h}$ of each other. The operator B_h is not a consistent approximation to the identity operator but satisfies

$$\|B_h\|_{\ell_2} \leq 3.7 \quad \text{and} \quad \|B_h^{-1}\|_{\ell_2} \leq 2.2 .$$

The other matrix, K_h, is an approximation to the correct compact operator and its singular values converge to those of the continuous operator. These results suffice to explain the rapid convergence which results when the capacitance matrix equation is solved by a conjugate gradient method, see Hayes [16], Proskurowski and Widlund [31], Shieh [33] and Section 4 of this paper.

4. Four capacitance matrix algorithms. When we start considering the implementation of these ideas, we are immediately faced with the question of generating and storing the capacitance matrix C. This matrix is a dense nonsymmetric, $p \times p$ matrix, where p is the number of irregular mesh points. Many problems in the plane can be satisfactorily solved using a value of p less than 200 and, in such cases, we can normally afford to deal with C explicitly. For problems in three dimensions, we obtain values of p in excess of 1000 already for very coarse meshes and a method similar to that of George [15] must be used, see discussion below. We note that if N is the total number of mesh points p grows in proportion to $N^{1/2}$ and $N^{2/3}$ in two and three dimensions respectively. The cost of using a highly accurate Helmholtz solver, based on fast Fourier transforms, is on the order of const. $N \log_2 N$. The constant is less than or equal to three if the mesh is well chosen. A well written program of this type requires only $N + o(N)$ storage locations.

We will now describe how the capacitance matrix can be generated. Because of the sparsity of the matrices U, Z and V, we see that we need only on the order of p^2 elements of B^{-1} to compute the matrix C. If periodic boundary conditions are chosen the matrix B^{-1} will be a circulant and we have complete information on B^{-1} if we know one of its columns. To see this, we note that because of periodicity the potential at one mesh point, due to a unit charge at another mesh point, depends only on the difference of the coordinates of the two points. A column of B^{-1} is available at the expense of one call of the fast solver subroutine and the calculation of C can then be completed at the additional expense of const. p^2 arithmetic operations.

We are now ready to describe the first variant of our capacitance matrix method. We first generate the capacitance matrix C as indicated above. We then apply a Gaussian elimination subroutine storing the triangular factors in the array which contained C. This part of the program requires $p^3/3$ multiplications and it clearly dominates the execution time for large values of p. For each set of data we then need p^2 multiplications to solve the capacitance matrix equation and two calls of the fast solver subroutine. This is a very efficient method if many sets of data have to be handled and we can afford to store C.

In our second variant of the method no Gaussian elimination is carried out. Instead the capacitance matrix equation, which is of the form (3.1) for the Dirichlet case, is solved by a conjugate gradient method, see Hestenes [17], Hestenes and Stiefel [18], Lanczos [24] and Proskurowski and Widlund [31]. Since C is nonsymmetric, we reformulate the problem in terms of a least squares problem. Each step of the conjugate gradient method requires the calculation of $C^T(Cx)$, where x is a given vector, and therefore requires $2p^2 + o(p^2)$ multiplications. Because of a favorable distribution of the singular values of C, 12-20 iterations will suffice, for a large family of problems in the plane, in order to obtain 5-6 correct decimal digits in the solution of the discrete Helmholtz problem, see further Proskurowski and Widlund [31] and Proskurowski [30]. This method is faster than the first for a single set of data and moderately large values of p. Programs implementing these two variants are listed in the ERDA-NYU report by Proskurowski and Widlund [31].

We will now discuss more recent work, see further O'Leary and Widlund [26] and Proskurowski [30], in which methods are developed which do not require C explicitly. If we concentrate our discussion on the Dirichlet case, we see from equation (3.1) that we can compute Cx as follows. Generate the mesh function Vx, use the fast solver to obtain $B^{-1}Vx$ and compute $Z^TB^{-1}Vx$ at an expense on the order of p operations. The vector C^TCx can be obtained in this fashion at a cost of essentially two calls of the fast solver subroutine. The conjugate gradient method is also used in this third variant of the method.

In the code developed by O'Leary for the three dimensional case 4 integer and 7 real arrays of dimension p are used. In addition two three dimensional arrays are used for mesh functions on the entire mesh. One of these arrays is not needed if the right-hand side, f, of the

Helmholtz equation vanishes. The arrays of dimension p are used to carry all geometric information, the Dirichlet data and as work space for the conjugate gradient iteration. The boundary is described by the coordinates of the irregular points and the distances, with sign, from these points to the boundary along the coordinate directions. The program allows for the possibility that an irregular point can have more than one exterior neighbor in a coordinate direction. We note that the distances to the boundary from the irregular points are needed in order to obtain a better than first order accurate difference approximation. The approximate information on the direction of the normal to the boundary, which can be obtained from this data, suffices for the construction of discrete dipoles. The program which will soon be released, also contains a subroutine which checks, among other things, that no boundary points are missing. Proskurowski has recently developed a similar program for problems in two dimensions.

In our fourth variant, we seek to eliminate the need for the large arrays, of dimension N, used in the methods described so far. A fast Helmholtz solver, using only a fraction of this number of storage locations, has been designed. The data must be accessed twice, but the amount of arithmetic is less than two times that needed for the corresponding conventional fast solver. This solver is to be used only to generate the right-hand side of the capacitance matrix equation and the final solution. The iteration which determines the discrete single or dipole density is carried out as in the third variant, except that still another fast Helmholtz solver is used. This method, developed by Banegas [1], exploits the fact that the data for these Helmholtz problems differ from zero only at a number of points which is on the order of p and that similarly the solutions are needed only in a neighborhood of the points of $\partial\Omega_h$. A Fortran program has been developed by Proskurowski [30] for problems in two dimensions. If the data is given in terms of a Fortran function, we will be able to solve problems with several hundred thousand variables, entirely in core, using less than one hundred thousand storage locations.

Early numerical experiments are described in detail in Proskurowski and Widlund [31]. We will now give some preliminary results from more recent experiments.

A Dirichlet problem for a sphere, and with $c = 0$, was solved by our code for three dimensional problems. This program implements the third method. We used 24576 mesh points out of which 8796 belonged to the sphere and 1698 were irregular. In 13 iterations between 4 and 5 correct decimal digits were obtained. The CPU time on the CDC 6600 of the

Courant Institute, using a FTN compiler, was 166.1 seconds out of which 91% was used up by 28 calls of the Helmholtz solver on the cube. Including the program, 65000 words of storage were required for this run. Our program can probably be speeded up noticeably by upgrading the fast Fourier transform. In a comparison with a smaller problem, we noticed an increase of the CPU time by a factor 5.95 when the number of variables was increased by a factor 6.48.

We finally give the CPU time required for a problem in two dimensions. The region is a circle with 49 mesh points across. We have 1921 unknowns and 132 irregular mesh points. A CDC 7600 at the Lawrence Berkeley Laboratory using a FTN4, OPT = 2, compiler was used. Between 5 and 6 decimal digits were obtained for variants 2-4, a much higher accuracy for variant 1.

CPU time in seconds for a problem on a circle

	Variant 1	Variant 2	Variant 3	Variant 4
Generation of C	0.475	0.475	-	-
Factorization of C	0.602	-	-	-
Number of iterations	-	16	12	13
Total execution time	1.215	0.944	1.853	1.472
Time to solve an additional problem	0.138	0.469	1.853	1.440

References

1. Banegas, A., "Fast Poisson Solvers for Sparse Data," to appear.

2. Bank, R.E., "Marching Algorithms and Gaussian Elimination," Proc. Symp. on Sparse Matrix Computations, Argonne National Lab., Sept. 1975, Edited by J.R. Bunch and D.J. Rose, Academic Press.

3. Bartels, R. and Daniel, J.W., "A Conjugate Gradient Approach to Nonlinear Elliptic Boundary Value Problems in Irregular Regions," Conference on the Numerical Solution of Differential Equations, Dundee, Scotland, July 1973, Lecture Notes in Mathematics, Springer, Vol. 363, pp. 1-11.

4. Buneman, O., "A Compact Non-Iterative Poisson Solver," Rep. SUIPR-294, Inst. Plasma Research, Stanford University, 1969.

5. Buzbee, B.L., "A Capacitance Matrix Technique," Proc. Symp. on Sparse Matrix Computations, Argonne National Lab., Sept. 1975, Edited by J.R. Bunch and D.J. Rose, Academic Press.

6. Buzbee, B.L. and Dorr, F.W., "The Direct Solution of the Biharmonic Equation on Rectangular Regions and the Poisson Equation on Irregular Regions," SIAM J. Numer. Anal., Vol. 11, 1974, pp. 753-763.

7. Buzbee, B.L., Dorr, F.W., George, J.A. and Golub, G.H., "The Direct Solution of the Discrete Poisson Equation on Irregular Regions," SIAM J. Numer. Anal., Vol. 8, 1971, pp. 722-736.

8. Buzbee, B.L., Golub, G.H. and Nielson, C.W., "On Direct Methods for Solving Poisson's Equation," SIAM J. Numer. Anal., Vol. 7, 1970, pp. 627-656.

9. Collatz, L., "The Numerical Treatment of Differential Equations," Springer, 1966.

10. Concus, P. and Golub, G.H., "A Generalized Conjugate Gradient Method for Nonsymmetric Systems of Linear Equations," Proc. 2nd Int. Symp. on Computing Methods in Applied Sciences and Engineering, IRIA, Paris, Dec. 1975, to appear.

11. Concus, P., Golub, G.H. and O'Leary, D.P., "A Generalized Conjugate Gradient Method for the Numerical Solution of Elliptic Partial Differential Equations," Proc. Symp. on Sparse Matrix Computations, Argonne National Lab., Sept. 1975. Edited by J.R. Bunch and D.J. Rose, Academic Press.

12. Courant, R. and Hilbert, D., "Methods of Mathematical Physics," Interscience, 1953.

13. Fischer, D., Golub, G., Hald, O., Leiva, C. and Widlund, O., "On Fourier-Toeplitz Methods for Separable Elliptic Problems," Math. Comp., Vol. 28, 1974, pp. 349-368.

14. Garabedian P.R., "Partial Differential Equations," Wiley, 1964.

15. George, J.A., "The Use of Direct Methods for the Solution of the Discrete Poisson Equation on Non-Rectangular Regions," Computer Science Department Report 159, Stanford University, 1970.

16. Hayes, R.M., "Iterative Methods of Solving Linear Problems on Hilbert Space," Contributions to the Solution of Systems of Linear Equations and the Determination of Eigenvalues. Ed. by O. Taussky. Nat. Bur. of Standards Applied Math. Series, Vol. 39, 1954, pp. 71-103.

17. Hestenes, M. R., "The Conjugate Gradient Method for Solving Linear Systems," Proc. Symp. Appl. Math. VI, Numer. Anal., 1956, pp. 83-102.

18. Hestenes, M.R. and Stiefel, E., "Method of Conjugate Gradients for Solving Linear Systems," J. Res. Nat. Bur. Standards, Vol. 49, 1952, pp. 409-436.

19. Hockney, R.W., "A Fast Direct Solution of Poisson's Equation Using Fourier Analysis," J. Assoc. Comp. Mach., Vol. 12, 1965, pp. 95-113.

20. Hockney, R.W., "Formation and Stability of Virtual Electrodes in a Cylinder," J. Appl. Phys., Vol. 39, 1968, pp. 4166-4170.

21. Hockney, R.W., "The Potential Calculation and Some Applications," Methods in Computational Physics, Vol. 9, 1970, Academic Press.

22. Hockney, R.W., "POT 4 - A Fast Direct Poisson Solver for the Rectangle Allowing Some Mixed Boundary Conditions and Internal Electrodes," IBM Research, R.C. 2870, 1970.

23. Householder, A.S., "The Theory of Matrices in Numerical Analysis," Blaisdell, 1964.

24. Lanczos, C., "Solution of Systems of Linear Equations by Minimized Iterations," J. Res. Nat. Bur. Standards, Vol. 49, 1952, pp. 33-53.

25. O'Leary, D.P., "Hybrid Conjugate Gradient Algorithms for Elliptic Systems," Computer Science Dept., Report 548, Stanford University, 1976.

26. O'Leary, D.P. and Widlund, O., ERDA-NYU report, to appear.

27. Pereyra, V., Proskurowski, W. and Widlund, O., "High Order Fast Laplace Solvers for the Dirichlet Problem on General Regions," Math. Comp., to appear.

28. Petrowsky, I.G., "Partial Differential Equations," Interscience, 1954.

29. Proskurowski, W., "On the Numerical Solution of the Eigenvalue Problem of the Laplace Operator by the Capacitance Matrix Method," Computer Science Dept. Report TRITA-NA-7609, Royal Institute of Technology, Stockholm, Sweden.

30. Proskurowski, W., Lawrence Berkeley Laboratory report, to appear.

31. Proskurowski, W. and Widlund, O., "On the Numerical Solution of Helmholtz's Equation by the Capacitance Matrix Method," Math. Comp., Vol. 30, 1976, pp. 433-468. Appeared also as an ERDA-NYU report COO-3077-99.

32. Schröder, J. and Trottenberg, U., "Reduktionsverfahren für Differenzengleichungen bei Randwertaufgaben I," Numer. Math., Vol. 22, 1973, pp. 37-68.

33. Shieh, A., "Fast Poisson Solver on Nonrectangular Domains," New York University Ph.D. thesis.

34. Swarztrauber, P. and Sweet, R., "Efficient FORTRAN Subprograms for the Solution of Elliptic Partial Differential Equations," Report, NCAR-TN/1A-109, National Center for Atmospheric Research, Boulder, Colorado, 1975.

35. Widlund, O., "On the Use of Fast Methods for Separable Finite Difference Equations for the Solution of General Elliptic Problems," Sparse Matrices and Their Applications, Ed. by D.J. Rose and R.A. Willoughby, Plenum Press, 1972.

36. Widlund, O., "A Lanczos Method for a Class of Non-Symmetric Systems of Linear Equations," to appear.

Vol. 460: O. Loos, Jordan Pairs. XVI, 218 pages. 1975.

Vol. 461: Computational Mechanics. Proceedings 1974. Edited by J. T. Oden. VII, 328 pages. 1975.

Vol. 462: P. Gérardin, Construction de Séries Discrètes p-adiques. »Sur les séries discrètes non ramifiées des groupes réductifs déployés p-adiques«. III, 180 pages. 1975.

Vol. 463: H.-H. Kuo, Gaussian Measures in Banach Spaces. VI, 224 pages. 1975.

Vol. 464: C. Rockland, Hypoellipticity and Eigenvalue Asymptotics. III, 171 pages. 1975.

Vol. 465: Séminaire de Probabilités IX. Proceedings 1973/74. Edité par P. A. Meyer. IV, 589 pages. 1975.

Vol. 466: Non-Commutative Harmonic Analysis. Proceedings 1974. Edited by J. Carmona, J. Dixmier and M. Vergne. VI, 231 pages. 1975.

Vol. 467: M. R. Essén, The Cos $\pi\lambda$ Theorem. With a paper by Christer Borell. VII, 112 pages. 1975.

Vol. 468: Dynamical Systems - Warwick 1974. Proceedings 1973/74. Edited by A. Manning. X, 405 pages. 1975.

Vol. 469: E. Binz, Continuous Convergence on C(X). IX, 140 pages. 1975.

Vol. 470: R. Bowen, Equilibrium States and the Ergodic Theory of Anosov Diffeomorphisms. III, 108 pages. 1975.

Vol. 471: R. S. Hamilton, Harmonic Maps of Manifolds with Boundary. III, 168 pages. 1975.

Vol. 472: Probability-Winter School. Proceedings 1975. Edited by Z. Ciesielski, K. Urbanik, and W. A. Woyczyński. VI, 283 pages. 1975.

Vol. 473: D. Burghelea, R. Lashof, and M. Rothenberg, Groups of Automorphisms of Manifolds. (with an appendix by E. Pedersen) VII, 156 pages. 1975.

Vol. 474: Séminaire Pierre Lelong (Analyse) Année 1973/74. Edité par P. Lelong. VI, 182 pages. 1975.

Vol. 475: Répartition Modulo 1. Actes du Colloque de Marseille-Luminy, 4 au 7 Juin 1974. Edité par G. Rauzy. V, 258 pages. 1975. 1975.

Vol. 476: Modular Functions of One Variable IV. Proceedings 1972. Edited by B. J. Birch and W. Kuyk. V, 151 pages. 1975.

Vol. 477: Optimization and Optimal Control. Proceedings 1974. Edited by R. Bulirsch, W. Oettli, and J. Stoer. VII, 294 pages. 1975.

Vol. 478: G. Schober, Univalent Functions – Selected Topics. V, 200 pages. 1975.

Vol. 479: S. D. Fisher and J. W. Jerome, Minimum Norm Extremals in Function Spaces. With Applications to Classical and Modern Analysis. VIII, 209 pages. 1975.

Vol. 480: X. M. Fernique, J. P. Conze et J. Gani, Ecole d'Eté de Probabilités de Saint-Flour IV-1974. Edité par P.-L. Hennequin. XI, 293 pages. 1975.

Vol. 481: M. de Guzmán, Differentiation of Integrals in R^n. XII, 226 pages. 1975.

Vol. 482: Fonctions de Plusieurs Variables Complexes II. Séminaire François Norguet 1974–1975. IX, 367 pages. 1975.

Vol. 483: R. D. M. Accola, Riemann Surfaces, Theta Functions, and Abelian Automorphisms Groups. III, 105 pages. 1975.

Vol. 484: Differential Topology and Geometry. Proceedings 1974. Edited by G. P. Joubert, R. P. Moussu, and R. H. Roussarie. IX, 287 pages. 1975.

Vol. 485: J. Diestel, Geometry of Banach Spaces – Selected Topics. XI, 282 pages. 1975.

Vol. 486: S. Stratila and D. Voiculescu, Representations of AF-Algebras and of the Group U (∞). IX, 169 pages. 1975.

Vol. 487: H. M. Reimann und T. Rychener, Funktionen beschränkter mittlerer Oszillation. VI, 141 Seiten. 1975.

Vol. 488: Representations of Algebras, Ottawa 1974. Proceedings 1974. Edited by V. Dlab and P. Gabriel. XII, 378 pages. 1975.

Vol. 489: J. Bair and R. Fourneau, Etude Géométrique des Espaces Vectoriels. Une Introduction. VII, 185 pages. 1975.

Vol. 490: The Geometry of Metric and Linear Spaces. Proceedings 1974. Edited by L. M. Kelly. X, 244 pages. 1975.

Vol. 491: K. A. Broughan, Invariants for Real-Generated Uniform Topological and Algebraic Categories. X, 197 pages. 1975.

Vol. 492: Infinitary Logic: In Memoriam Carol Karp. Edited by D. W. Kueker. VI, 206 pages. 1975.

Vol. 493: F. W. Kamber and P. Tondeur, Foliated Bundles and Characteristic Classes. XIII, 208 pages. 1975.

Vol. 494: A Cornea and G. Licea. Order and Potential Resolvent Families of Kernels. IV, 154 pages. 1975.

Vol. 495: A. Kerber, Representations of Permutation Groups II. V, 175 pages. 1975.

Vol. 496: L. H. Hodgkin and V. P. Snaith, Topics in K-Theory. Two Independent Contributions. III, 294 pages. 1975.

Vol. 497: Analyse Harmonique sur les Groupes de Lie. Proceedings 1973–75. Edité par P. Eymard et al. VI, 710 pages. 1975.

Vol. 498: Model Theory and Algebra. A Memorial Tribute to Abraham Robinson. Edited by D. H. Saracino and V. B. Weispfenning. X, 463 pages. 1975.

Vol. 499: Logic Conference, Kiel 1974. Proceedings. Edited by G. H. Müller, A. Oberschelp, and K. Potthoff. V, 651 pages 1975.

Vol. 500: Proof Theory Symposion, Kiel 1974. Proceedings. Edited by J. Diller and G. H. Müller. VIII, 383 pages. 1975.

Vol. 501: Spline Functions, Karlsruhe 1975. Proceedings. Edited by K. Böhmer, G. Meinardus, and W. Schempp. VI, 421 pages. 1976.

Vol. 502: János Galambos, Representations of Real Numbers by Infinite Series. VI, 146 pages. 1976.

Vol. 503: Applications of Methods of Functional Analysis to Problems in Mechanics. Proceedings 1975. Edited by P. Germain and B. Nayroles. XIX, 531 pages. 1976.

Vol. 504: S. Lang and H. F. Trotter, Frobenius Distributions in GL_2-Extensions. III, 274 pages. 1976.

Vol. 505: Advances in Complex Function Theory. Proceedings 1973/74. Edited by W. E. Kirwan and L. Zalcman. VIII, 203 pages. 1976.

Vol. 506: Numerical Analysis, Dundee 1975. Proceedings. Edited by G. A. Watson. X, 201 pages. 1976.

Vol. 507: M. C. Reed, Abstract Non-Linear Wave Equations. VI, 128 pages. 1976.

Vol. 508: E. Seneta, Regularly Varying Functions. V, 112 pages. 1976.

Vol. 509: D. E. Blair, Contact Manifolds in Riemannian Geometry. VI, 146 pages. 1976.

Vol. 510: V. Poènaru, Singularités C^∞ en Présence de Symétrie. V, 174 pages. 1976.

Vol. 511: Séminaire de Probabilités X. Proceedings 1974/75. Edité par P. A. Meyer. VI, 593 pages. 1976.

Vol. 512: Spaces of Analytic Functions, Kristiansand, Norway 1975. Proceedings. Edited by O. B. Bekken, B. K. Øksendal, and A. Stray. VIII, 204 pages. 1976.

Vol. 513: R. B. Warfield, Jr. Nilpotent Groups. VIII, 115 pages. 1976.

Vol. 514: Séminaire Bourbaki vol. 1974/75. Exposés 453 – 470. IV, 276 pages. 1976.

Vol. 515: Bäcklund Transformations. Nashville, Tennessee 1974. Proceedings. Edited by R. M. Miura. VIII, 295 pages. 1976.

Vol. 516: M. L. Silverstein, Boundary Theory for Symmetric Markov Processes. XVI, 314 pages. 1976.

Vol. 517: S. Glasner, Proximal Flows. VIII, 153 pages. 1976.

Vol. 518: Séminaire de Théorie du Potentiel, Proceedings Paris 1972–1974. Edité par F. Hirsch et G. Mokobodzki. VI, 275 pages. 1976.

Vol. 519: J. Schmets, Espaces de Fonctions Continues. XII, 150 pages. 1976.

Vol. 520: R. H. Farrell, Techniques of Multivariate Calculation. X, 337 pages. 1976.

Vol. 521: G. Cherlin, Model Theoretic Algebra - Selected Topics. IV, 234 pages. 1976.

Vol. 522: C. O. Bloom and N. D. Kazarinoff, Short Wave Radiation Problems in Inhomogeneous Media: Asymptotic Solutions. V. 104 pages. 1976.

Vol. 523: S. A. Albeverio and R. J. Høegh-Krohn, Mathematical Theory of Feynman Path Integrals. IV, 139 pages. 1976.

Vol. 524: Séminaire Pierre Lelong (Analyse) Année 1974/75. Edité par P. Lelong. V, 222 pages. 1976.

Vol. 525: Structural Stability, the Theory of Catastrophes, and Applications in the Sciences. Proceedings 1975. Edited by P. Hilton. VI, 408 pages. 1976.

Vol. 526: Probability in Banach Spaces. Proceedings 1975. Edited by A. Beck. VI, 290 pages. 1976.

Vol. 527: M. Denker, Ch. Grillenberger, and K. Sigmund, Ergodic Theory on Compact Spaces. IV, 360 pages. 1976.

Vol. 528: J. E. Humphreys, Ordinary and Modular Representations of Chevalley Groups. III, 127 pages. 1976.

Vol. 529: J. Grandell, Doubly Stochastic Poisson Processes. X, 234 pages. 1976.

Vol. 530: S. S. Gelbart, Weil's Representation and the Spectrum of the Metaplectic Group. VII, 140 pages. 1976.

Vol. 531: Y.-C. Wong, The Topology of Uniform Convergence on Order-Bounded Sets. VI, 163 pages. 1976.

Vol. 532: Théorie Ergodique. Proceedings 1973/1974. Edité par J.-P. Conze and M. S. Keane. VIII, 227 pages. 1976.

Vol. 533: F. R. Cohen, T. J. Lada, and J. P. May, The Homology of Iterated Loop Spaces. IX, 490 pages. 1976.

Vol. 534: C. Preston, Random Fields. V, 200 pages. 1976.

Vol. 535: Singularités d'Applications Differentiables. Plans-sur-Bex. 1975. Edité par O. Burlet et F. Ronga. V, 253 pages. 1976.

Vol. 536: W. M. Schmidt, Equations over Finite Fields. An Elementary Approach. IX, 267 pages. 1976.

Vol. 537: Set Theory and Hierarchy Theory. Bierutowice, Poland 1975. A Memorial Tribute to Andrzej Mostowski. Edited by W. Marek, M. Srebrny and A. Zarach. XIII, 345 pages. 1976.

Vol. 538: G. Fischer, Complex Analytic Geometry. VII, 201 pages. 1976.

Vol. 539: A. Badrikian, J. F. C. Kingman et J. Kuelbs, Ecole d'Eté de Probabilités de Saint Flour V-1975. Edité par P.-L. Hennequin. IX, 314 pages. 1976.

Vol. 540: Categorical Topology, Proceedings 1975. Edited by E. Binz and H. Herrlich. XV, 719 pages. 1976.

Vol. 541: Measure Theory, Oberwolfach 1975. Proceedings. Edited by A. Bellow and D. Kölzow. XIV, 430 pages. 1976.

Vol. 542: D. A. Edwards and H. M. Hastings, Čech and Steenrod Homotopy Theories with Applications to Geometric Topology. VII, 296 pages. 1976.

Vol. 543: Nonlinear Operators and the Calculus of Variations, Bruxelles 1975. Edited by J. P. Gossez, E. J. Lami Dozo, J. Mawhin, and L. Waelbroeck, VII, 237 pages. 1976.

Vol. 544: Robert P. Langlands, On the Functional Equations Satisfied by Eisenstein Series. VII, 337 pages. 1976.

Vol. 545: Noncommutative Ring Theory. Kent State 1975. Edited by J. H. Cozzens and F. L. Sandomierski. V, 212 pages. 1976.

Vol. 546: K. Mahler, Lectures on Transcendental Numbers. Edited and Completed by B. Diviš and W. J. Le Veque. XXI, 254 pages. 1976.

Vol. 547: A. Mukherjea and N. A. Tserpes, Measures on Topological Semigroups: Convolution Products and Random Walks. V, 197 pages. 1976.

Vol. 548: D. A. Hejhal, The Selberg Trace Formula for PSL (2, ℝ). Volume I. VI, 516 pages. 1976.

Vol. 549: Brauer Groups, Evanston 1975. Proceedings. Edited by D. Zelinsky. V, 187 pages. 1976.

Vol. 550: Proceedings of the Third Japan - USSR Symposium on Probability Theory. Edited by G. Maruyama and J. V. Prokhorov. VI, 722 pages. 1976.

Vol. 551: Algebraic K-Theory, Evanston 1976. Proceedings. Edited by M. R. Stein. XI, 409 pages. 1976.

Vol. 552: C. G. Gibson, K. Wirthmüller, A. A. du Plessis and E. J. N. Looijenga. Topological Stability of Smooth Mappings. V, 155 pages. 1976.

Vol. 553: M. Petrich, Categories of Algebraic Systems. Vector and Projective Spaces, Semigroups, Rings and Lattices. VIII, 217 pages. 1976.

Vol. 554: J. D. H. Smith, Mal'cev Varieties. VIII, 158 pages. 1976.

Vol. 555: M. Ishida, The Genus Fields of Algebraic Number Fields. VII, 116 pages. 1976.

Vol. 556: Approximation Theory. Bonn 1976. Proceedings. Edited by R. Schaback and K. Scherer. VII, 466 pages. 1976.

Vol. 557: W. Iberkleid and T. Petrie, Smooth S^1 Manifolds. III, 163 pages. 1976.

Vol. 558: B. Weisfeiler, On Construction and Identification of Graphs. XIV, 237 pages. 1976.

Vol. 559: J.-P. Caubet, Le Mouvement Brownien Relativiste. IX, 212 pages. 1976.

Vol. 560: Combinatorial Mathematics, IV, Proceedings 1975. Edited by L. R. A. Casse and W. D. Wallis. VII, 249 pages. 1976.

Vol. 561: Function Theoretic Methods for Partial Differential Equations. Darmstadt 1976. Proceedings. Edited by V. E. Meister, N. Weck and W. L. Wendland. XVIII, 520 pages. 1976.

Vol. 562: R. W. Goodman, Nilpotent Lie Groups: Structure and Applications to Analysis. X, 210 pages. 1976.

Vol. 563: Séminaire de Théorie du Potentiel. Paris, No. 2. Proceedings 1975-1976. Edited by F. Hirsch and G. Mokobodzki. VI, 292 pages. 1976.

Vol. 564: Ordinary and Partial Differential Equations, Dundee 1976. Proceedings. Edited by W. N. Everitt and B. D. Sleeman. XVIII, 551 pages. 1976.

Vol. 565: Turbulence and Navier Stokes Equations. Proceedings 1975. Edited by R. Temam. IX, 194 pages. 1976.

Vol. 566: Empirical Distributions and Processes. Oberwolfach 1976. Proceedings. Edited by P. Gaenssler and P. Révész. VII, 146 pages. 1976.

Vol. 567: Séminaire Bourbaki vol. 1975/76. Exposés 471-488. IV, 303 pages. 1977.

Vol. 568: R. E. Gaines and J. L. Mawhin, Coincidence Degree, and Nonlinear Differential Equations. V, 262 pages. 1977.

Vol. 569: Cohomologie Etale SGA 4½. Séminaire de Géométrie Algébrique du Bois-Marie. Edité par P. Deligne. V, 312 pages. 1977.

Vol. 570: Differential Geometrical Methods in Mathematical Physics, Bonn 1975. Proceedings. Edited by K. Bleuler and A. Reetz. VIII, 576 pages. 1977.

Vol. 571: Constructive Theory of Functions of Several Variables, Oberwolfach 1976. Proceedings. Edited by W. Schempp and K. Zeller. VI, 290 pages. 1977

Vol. 572: Sparse Matrix Techniques, Copenhagen 1976. Edited by V. A. Barker. V, 184 pages. 1977.

Vol. 573: Group Theory, Canberra 1975. Proceedings. Edited by R. A. Bryce, J. Cossey and M. F. Newman. VII, 146 pages. 1977.

Vol. 574: J. Moldestad, Computations in Higher Types. IV, 203 pages. 1977.

Vol. 575: K-Theory and Operator Algebras, Athens, Georgia 1975. Edited by B. B. Morrel and I. M. Singer. VI, 191 pages. 1977.

Vol. 576: V. S. Varadarajan, Harmonic Analysis on Real Reductive Groups. VI, 521 pages. 1977.

Vol. 577: J. P. May, E_∞ Ring Spaces and E_∞ Ring Spectra. IV, 268 pages. 1977.

Vol. 578: Séminaire Pierre Lelong (Analyse) Année 1975/76. Edité par P. Lelong. VI, 327 pages. 1977.

Vol. 579: Combinatoire et Représentation du Groupe Symétrique, Strasbourg 1976. Proceedings 1976. Edité par D. Foata. IV, 339 pages. 1977.